The Cortex and the Critical Point

The Cortex and the Critical Point

Understanding the Power of Emergence

John M. Beggs

The MIT Press
Cambridge, Massachusetts
London, England

The MIT Press would like to thank the anonymous peer reviewers who provided comments on drafts of this book. The generous work of academic experts is essential for establishing the authority and quality of our publications. We acknowledge with gratitude the contributions of these otherwise uncredited readers.

This book was set in Times New Roman by Westchester Publishing Services. Printed and bound in the United States of America.

Library of Congress Cataloging-in-Publication Data

Names: Beggs, John M., author.
Title: The cortex and the critical point : understanding the power of emergence /
   John M. Beggs.
Description: Cambridge, Massachusetts : The MIT Press, [2022] | Includes bibliographical
   references and index.
Identifiers: LCCN 2021046107 | ISBN 9780262544030 (paperback)
Subjects: LCSH: Cerebral cortex. | Brain. | Neurology.
Classification: LCC QP383 .B44 2022 | DDC 612.8/25—dc23/eng/20211109
LC record available at https://lccn.loc.gov/2021046107

10 9 8 7 6 5 4 3 2 1

This book is dedicated to my father, who gave me curiosity and an interest in science.

Contents

Acknowledgments

I would not have been able to finish this book without help from many others. It is fitting for me to acknowledge their contributions here.

First, I am grateful to Dietmar Plenz, who was my postdoctoral advisor when we discovered neuronal avalanches. He gave me the opportunity to work on a project analyzing activity from multielectrode arrays. Neither of us knew what that would bring, and the three years I spent in his lab were filled with amazement at what those data produced. Dietmar is an excellent scientist, and I learned from him how to be more precise and rigorous. He is a dynamic investigator and unafraid to tackle the most challenging problems. Thank you for teaching me so much.

Second, to my colleagues who reviewed early drafts of these chapters. To Gerardo Ortiz, a condensed matter theorist who has educated me tremendously in critical phenomena. Thank you, my long-time friend for keeping me honest about the physics. To Ehren Newman, a systems neuroscientist who helped me to make the material more relevant and understandable. Thank you for not being afraid to tell me the truth. "Wounds from a friend can be trusted, but an enemy multiplies kisses" (Proverbs 27:6). To Katy Hagen, a graduate student and writer who made the text more readable. Thank you for reading every sentence and every comma with a helpful and critical eye. You all have made the book immeasurably better, but the errors that remain are mine alone.

Third, to my students, who have really done all the hard work on our criticality projects: Rashid Williams-Garcia, Mark Moore, Shinya Ito, Nick Timme, Najja Marshall, and Leandro Fosque; and to Masanori Shimono who was a superb postdoc.

Fourth, to Karin Dahmen and Tom Butler, for suggesting we look at the exponent relation and avalanche shape collapse. You have been excellent collaborators and fun to work with.

Fifth, to my wonderful collaborators David Hsu and Alan Litke. David introduced me to the possible connections between supercriticality and epilepsy. Alan developed a cutting-edge electrode array that made high-resolution recordings of neuronal avalanches possible.

Sixth, to Bob Prior and Olaf Sporns, for believing in me and encouraging me to write this book. It has been as gratifying to do this as you said it would be.

Seventh, to the National Science Foundation who has generously funded our lab over so many years. I am grateful to Ken Whang and Krastan Blagoev who supported much of the work described in these chapters.

Eighth, to my family. To Sara my wife for mowing the lawn, planning our vacations, and doing everything else while I was in a temporary vegetative state writing this book. You truly love me by grace and not based on my works! To my daughters Zoe and Katerina, who missed out on our regular coffee and donuts while I finished these chapters. Completing this book gives me great joy, but it only comes once. This does not compare to what I have in you every day of my life.

Last but not least, I thank God for making nature fascinating and for leading me through a path where I have seen its wonders. At every step, you have provided.

Introduction

The Brain—is wider than the Sky—
For—put them side by side—
The one the other will contain
With ease—and you—beside—

The Brain is deeper than the sea—
For—hold them—Blue to Blue—
The one the other will absorb—
As sponges—Buckets—do—
—Emily Dickinson, c. 1862

Individual neurons have limited computational powers, but when they work together, they are astonishingly brilliant. Figuring out *how* they work together is the most important task in understanding how the brain works. And understanding how the brain works is, as Emily Dickinson might argue, the question that contains them all.

I want to tell you something startling, almost magical, about how neurons work together. It is not easy to understand. In fact, it will take a whole book to fully explain. But if I were forced to try in just a paragraph, I would say this—it is like when water, at just the right pressure, changes into steam. For a moment it is both a flowing liquid and individual molecules zipping around through the air. Neurons can act that way too, firing synchronously and then breaking off to improvise by themselves. Just at this transition, they are paradoxically both independent and interdependent with all other neurons. Right here, near what we will call the critical point, information flows easily, computations are most facile, and the brain is exquisitely sensitive to inputs. Here, intricate patterns of waves, oscillations, and avalanches of activity arise most readily. Slip too far below this point, and neurons fall into the abyss of silence. Nudge above it, and they get swept up into the fatal storm of seizures. Right around the critical point there is a narrow passage that opens to an expanse of complexity and emergence that is wider than the sky and deeper than the sea.

What is this critical point, and why does it have these interesting properties? Perhaps the simplest way to understand the critical point is by considering the three ways activity could propagate in your brain. First, it could be damped, so that incoming signals quickly die out. You smell a rose, but it triggers no memory and no succeeding thoughts. Second, inputs could be amplified so that they rapidly grow. You smell a rose, and you think of your spouse and your first date, and then your wedding, and then what the kids were like when they were young, and you quickly wonder how you will pay for the kids' college tuition, and your thoughts accelerate until your saturated brain is overwhelmed with a seizure. The third way—the critical way—preserves the strength of inputs, neither weakening them nor strengthening them. Rather, when the network is near the critical point, it transforms incoming signals into different patterns that still preserve, as much as possible, their original information content. These signals can then bounce around within the brain a long time before dying out, colliding with other inputs, causing new patterns to form. You smell a rose, and you write a poem that connects your first date with your anniversary dinner.

It does not seem that we are born with brains right at the critical point. Research suggests that during development, sprouting connections compete with pruning to lead us to it (Tetzlaff et al. 2010; Stewart and Plenz 2008). Once near the critical point, there are homeostatic processes to keep us there. Lack of sleep seems to move us toward being overamplified, and seizures are known to occur more often in sleep-deprived people. But a good night's sleep reduces this amplification and brings the brain closer to the critical point (Meisel et al. 2013). Just like a thermostat keeping a room near a set temperature, the brain returns toward the critical point after being pushed away from it (Ma et al. 2019). The benefit of being near the critical point is that many information processing functions are thought to be optimized there (Shew and Plenz 2013).

But before we delve too deeply into the questions surrounding the critical point, let's first step back and take a broader view of current neuroscience research. This may help to put the idea of the critical point into its proper context.

The Critical Point in Context

This is, without doubt, a golden age for neuroscience. The last 30 years have seen an efflorescence of tools to manipulate the brain and collect data from it. It is now possible to record activity from every neuron in a zebrafish larva's brain while it is freely swimming and responding to stimuli (Kim et al. 2017). We have a nearly complete map of one hemisphere of the fly brain, with every neuron and most of its synapses accounted for (Pipkin 2020). The neurons responsible for a mouse's memory of an event can be recorded, tagged, and replayed by laser stimulation, causing the mouse to behave as if the event had happened again (Carrillo-Reid et al. 2019; Ramirez et al. 2013). Given this progress, it is reasonable to expect that in a decade or two we will record and stimulate most neurons in the mouse cortex (and maybe the monkey's) while it is interacting in a virtual reality environment, controlled by experimenters.

Once we reach these heights, will we then understand how the brain works? One might think so, but some of those working on the front lines don't share this optimism. In a recent article in *Nautilus* magazine (Guitchounts 2020), Harvard graduate student Grigori Guitchounts asks Professor Jeff Lichtman, who developed the brainbow technique for mapping the brain's connectome:

"I think the word 'understanding' has to undergo an evolution," Lichtman said, as we sat around his desk. "Most of us know what we mean when we say "I understand something." It makes sense to us. We can hold the idea in our heads. We can explain it with language. But if I asked, Do you understand New York City?" you would probably respond, "What do you mean?" There's all this complexity. If you can't understand New York City, it's not because you can't get access to the data. It's just there's so much going on at the same time. That's what a human brain is. It's millions of things happening simultaneously among different types of cells, neuromodulators, genetic components, things from the outside. There's no point when you can suddenly say, "I now understand the brain," just as you wouldn't say, "I now get New York City."

Yes, the details are immense and overwhelming, but must we give up on the project of understanding, or settle for some diminished version of it? Understanding necessarily means distilling general principles to the point where some details do not matter.

For comparison, let us look at what happened in celestial mechanics. At first there was the massive data collection with scientists squinting at night to track planets; some took decades to orbit the sun, so the work took generations. Early models hardly provided understanding and were little more than fitting the data. Tycho Brahe thought each planet moved on a circle around the sun. Any backward, or retrograde, motion could be accounted for by having a smaller circle, rotating the other way, on top of the main circle. If need be, stacking three or four such epicycles could fit any planet's motion. Later, Kepler thought ellipses worked better and this led him to discern three laws that described planetary motion. This set the stage for Newton's grand synthesis of universal gravitation, where all was governed by a single equation. It worked for the planets, no matter their distance or size. After only a few observations, Gauss relied on it to declare where the planetoid Ceres would reappear nine months later (Teets and Whitehead 1999). This type of understanding made us think the universe was predictable like a clockwork and gave us ambitions to master it. It was this confidence that drove technology, transforming the next centuries. Understanding our brains would be no less momentous.

A nice story, but could anything like this happen in neuroscience? Biology is messy and not like physics in that way. Perhaps there isn't yet a set of equations that could fit on a T-shirt to describe the brain, or even part of it, with any precision. But it does seem like we should at least try to move toward this. History says we must go there eventually. Indeed, there are several overarching theories of how the brain works already in the literature (Friston 2010; Carandini and Heeger 2012; Poggio 1990; Kelso, Dumas, and Tognoli 2013). Some of these are speculative and not widely embraced. Neuroscientists are cautious and want to see testable predictions with details.

It is into this milieu that I am offering the idea that the cortex operates in the vicinity of a critical point—*the criticality hypothesis* (Beggs 2008). It is not my idea alone; many people have contributed to it before me (Kauffman 1969; Wilson and Cowan 1972; Kelso 1984; Freeman 1987; Dunkelmann and Radons 1994; Bienenstock 1995; Herz and Hopfield 1995; Bak 1996; Chialvo and Bak 1999; De Carvalho and Prado 2000; Linkenkaer-Hansen et al. 2001; Greenfield and Lecar 2001; Worrell et al. 2002; Eurich, Herrmann, and Ernst 2002). You can judge for yourself whether it accounts for the data, generates specific testable predictions, and has general principles that transcend details. You can also decide if it is a step toward understanding.

Let us now discuss some of the background behind this idea. Over the last twenty years, there has been a growing body of research investigating this hypothesis. While this interest began with behavioral work and modeling studies decades ago, it grew most rapidly in the early 2000s, when experimental support for a critical point in the cortex first appeared. The criticality hypothesis is provocative because it claims to be a unifying framework at a time when neuroscience is dominated by data-driven work. It imports theory, not methods, from physics to understand the brain. If correct, the concept of a critical point could explain how information transmission, storage, computation, and controllability are simultaneously optimized in cortical networks. It also offers testable predictions about how failures to attain optimality could lead to neuropathologies like epilepsy.

The growth of this field is evidenced by steadily increasing citations, conferences focused on criticality, a breadth of new research approaches and consistent media attention. While some of the earliest work began with cortical slices in rats, investigations of the critical point now include diverse species like worms, zebrafish, turtles, mice, monkeys and humans. The techniques employed have spanned from electrode recordings to calcium imaging, electroencephalography (EEG), magnetoencephalography (MEG), and functional magnetic resonance imaging (fMRI). More recent work has shown the critical point is behaviorally relevant to sensory processing, is predictive of health outcomes, and that deviations from the critical point can be restored by sleep. Citing this research as inspiration, electrical engineers have found networks of memristors can self-organize to operate near the critical point to perform computations and learn (Stieg et al. 2012; Pike et al. 2020; Hochstetter et al. 2021). *New Scientist* magazine identified the hypothesis that the brain operates near the critical point as one of the top five mathematical ideas driving brain research (Barras 2013).

Despite the potential promise of this idea, there is still a lack of consensus among neuroscientists as to its validity. What could be causing this hesitation? One reason may be rooted in the evidence that was originally offered in support of the critical point. Early work relied heavily on the existence of power laws. Skeptics rightly pointed out that while power laws are necessary to establish the existence of a critical point, they are not sufficient to do so. Other mechanisms can also produce power laws, yet do not necessarily indicate a critical point. This led to several years of confusion. While the field ultimately responded with better methods for quantifying proximity to the critical point, time had gone by. In addition, these new methods are highly technical and clear descriptions of why they work have been rare. There is now a great need to clearly explain how these early objections have been resolved and how solid evidence supporting operation near the critical point has accumulated.

The Goals and Structure of This Book

For that reason, the main goal of this book is to explain the critical point and its relevance to the brain as clearly as possible. I describe the central concepts from an intuitive perspective first, with more detailed descriptions later, leaving most technical aspects for the appendix. I also use figures abundantly so readers will have the chance to grasp what is going on both verbally and visually. Many of these figures were generated by computer models that can be freely accessed through links in the appendix. I hope that curious readers will find them useful for their own explorations, and that tired professors may find them helpful in creating homework assignments. At the end of many chapters, there are suggested exercises. In addition, we have collected several example datasets so that students can perform analyses and test these ideas for themselves. I think that interacting with the material is crucial for understanding it.

Because this book is positioned as an introduction to the field, it is not a comprehensive review of criticality research. For that, there are already excellent books (Plenz and Niebur 2014; Tomen, Herrmann, and Ernst 2019) covering most aspects of current work with technical details. These can be read with great profit by those already in this area. Instead, this book is organized around using a few examples of models and data to convey some key principles. With respect to models, I have chosen to rely on the simple branching model, even though there are hundreds of different models in this area. With respect to data, I have emphasized spike recordings, even though the field routinely deals with all data types. In making these decisions I aimed to keep the presentation as clear and consistent as possible. It was simply not feasible in an introductory work to give each of the contributions in this area the detailed attention it deserves. I hope that by drawing attention to this field I will encourage readers to later learn for themselves about the research of the many excellent scientists I was not able to include here.

My own research for the past 20 years has been deeply involved with criticality. I co-authored a paper with Dietmar Plenz that contributed to spurring the growth of this field (Beggs and Plenz, 2003). In other work, my colleagues and I developed the critical branching model that has become a workhorse in studies of critical neural networks. This model illustrates how information transmission and information storage would be optimized simultaneously at the critical point (Haldeman and Beggs 2005). We also introduced the exponent relation as an authentic signature of proximity to the critical point in neuronal data (Friedman et al. 2012), an approach that is increasingly adopted. Most recently, we advanced a new theory, *quasicriticality*, about how cortical networks hover around the critical point (Williams-Garcia et al. 2014; Fosque et al. 2021), and this has received popular coverage (Ouellette 2018; Helias 2021). These experiences give me a perspective from which to describe the overall trajectory of this field, the challenges it has faced, and the potential promise it holds.

A key theme of this book is that neurons are far more powerful when they work together than when they operate independently. While this seems obvious, the focus on emergent properties in current neuroscience is still relatively underdeveloped. There has been excellent work on rhythms, synchrony, and waves, but more complex emergent phenomena are rarely discussed. With the current abundance of multichannel neuronal data, now is an ideal time to demonstrate that the framework of emergence has the potential to reveal still higher-level phenomena in the brain. The avalanches of neuronal activity that appear near the critical point are an emergent phenomenon that is more complex than those previously studied yet are still approachable when clearly explained. I hope to use the concept of the critical point and the cortex to illustrate the power of the emergent perspective, making it more accessible to a broad range of investigators.

Very broadly, the book has three parts. Part I is introductory and gives background to the main ideas behind the criticality hypothesis and emergent phenomena. Part II addresses the critical point and its main consequences. Part III explores issues that are likely to drive future research in this field. Here is a slightly more detailed overview of the book's structure:

Chapter 1 attempts to show readers, as briefly and clearly as possible, the central ideas underlying criticality research in neuroscience. Chapter 2 is a general discussion of emergent phenomena and how they can be understood. Here I employ some simple computer models for illustrations. This is meant to pave the way for a detailed discussion of the properties that emerge in networks near the critical point.

Part II is the heart of the book, dealing with the most central issues. In chapter 3, we go over the specific signatures of the critical point that can be seen in neural network models, and then we turn to the data to see if these exist. Once we have considered the evidence that the cortex operates near the critical point, we turn to the two main consequences of being there. The first consequence is scale-free properties that confer optimum information processing; this is covered in chapter 4. The second consequence is universality, covered in chapter 5. Universality is the idea that complex emergent behavior, like that seen near the critical point, can be explained by relatively simple models that are applicable across species and scales. If true, universality would show us that there is hope for understanding the brain without having to first know all the details. This concludes part II of the book.

In part III, we consider the future of this field. Chapter 6 covers how operation near the critical point is homeostatically regulated, like blood pressure and body temperature in healthy individuals. A consequence of this is that departures from the critical point are associated with neurological disorders. In chapter 7 we address quasicriticality, the idea that the brain can never fully reach the critical point because it is always being driven by inputs and noise. Very recent work supports this idea; we discuss this and competing theories as an issue not yet resolved. Chapter 8 focuses on how operating near the critical point allowed the cortex to amplify our intelligence. Chapter 9 is an epilogue, to briefly cover what we know, what we do not know, and remaining open questions.

The appendix contains technical descriptions that would have interrupted the flow of the text in the main chapters.

I have not yet explained in much detail what the criticality hypothesis is, or why operating near the critical point would profoundly affect a network. To begin that, let us move on to chapter 1, next.

I

Background

1

The Main Idea

It can scarcely be denied that the supreme goal of all theory is to make the irreducible basic elements as simple and as few as possible without having to surrender the adequate representation of a single datum of experience.
—Albert Einstein

I remember the first time I tried to tell people about the material in this chapter. It was at a conference with tens of thousands of people—I was standing in front of my poster, probably looking a little too eager, even desperate. I probably spoke too fast and too loudly, and while they nodded along with me, they looked puzzled and trickled away. Dejected, I thought I would dial back the enthusiasm and not try to play to the crowd. I slowed down and focused on one person at a time, walking them through what we found. It was deliberate and less dramatic, but they didn't seem to force their smiles at the end. Some even thanked me.

In this chapter, I want to capture some of that spirit of explaining things clearly. I first want to sketch the main idea of the critical point and its implications for information processing and the emergence of complexity. This sketch will use a conceptual model designed to convey intuitions; it therefore must necessarily be incomplete and lack some details. However, my hope is that it will explain why the critical point is interesting and will serve as an introduction for more detailed explanations later.

A Simple Model

Suppose we have neurons that can be either on (1) or off (0). When a neuron is on, its connections will determine how many other neurons, on average, it will activate next. Let's say weak connections typically activate less than one neuron, medium connections usually activate one, and strong connections usually activate two (figure 1.1).

Now let's consider a network of these neurons connected to each other. How activity spreads in the network is controlled by the connection strengths. With weak connection strengths, activating one neuron will usually lead to no activity at all in a short time. With strong connection strengths, activating one neuron will quickly lead to an explosion of activity. Of course, there is a middle ground where the connections can be given just the right strength, and where activating one neuron will on average lead to only one other

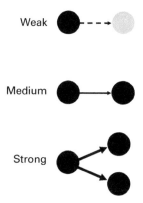

Figure 1.1
Connection strengths determine the average number of neurons activated. Each active neuron (black) has a probability of activating other neurons. Weak connections on average activate less than one neuron (gray); medium connections, one neuron; strong connections, two neurons.

neuron activated later. When the connection strengths are tuned to this level, the network is operating at the critical point (figure 1.2). While it is easy to define this critical point, it is more difficult to fathom its consequences. Even slight deviations from it over time will be amplified. As we will see, it is the Strait of Messina, poised between the Scylla of inactivity and the Charybdis of overactivity. The implications are varied and far-reaching.

Optimal Information Processing

How would being at this critical point affect information that could pass through the network? To understand this, let us frame information transmission through the network in terms of a guessing game. In this game, we will activate some randomly chosen number of neurons and let the network run for several time steps. Our job will be to look at the pattern of activity in the network at that later time and guess how many neurons were activated at the start. The better we can guess, the more information the network has transmitted about what its input was. Let's see how tuning the connection strengths affects this.

When the connections are weak, no matter how many neurons we activate at the beginning, after several time steps there is usually no activity at all. This makes guessing very difficult, because there is no trace left. In contrast, when the connections are strong, even if only one neuron is activated at the start, activity is amplified and nearly all neurons are on after a few time steps. Again, guessing is difficult, but now for a different reason: the network is too active. With nearly all the neurons on, it is hard to reconstruct whether one or five neurons were active at the start. But when the connection strengths are tuned to the critical intermediate value, we have the best chance of guessing what the input was at the start. If three neurons were activated, they will on average activate three others in each time step, again and again.

The number of correct guesses is an intuitive way of describing how much information the network has transmitted about the starting conditions.[1] If we plot the amount of information against the strength of the network connections, we see a curve with a peak in the middle (figure 1.3). The peak occurs at medium connection strength and shows that information transmission is maximized at the critical point.

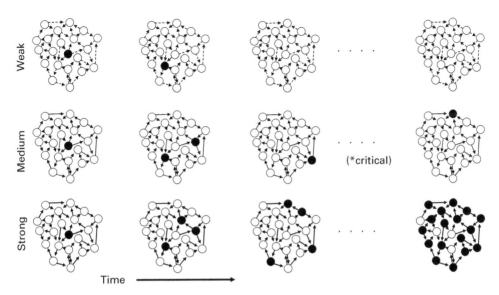

Figure 1.2
Connection strengths affect how activity propagates in a neural network. A single neuron is activated (black) in each recurrent network, and activity is transmitted with probability given by connection strength. Inactive neurons are white circles. Network activity unfolds over time toward the right. *Upper row*, Weak connections cause activity to quickly die out. *Middle row*, Medium connections approximately preserve the number of active neurons. This is the critical condition. *Bottom row*, Strong connections amplify activity, saturating the network.

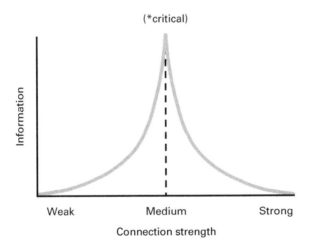

Figure 1.3
Information transmission is maximized near the critical point. Information is given by the ability to correctly guess, after several time steps, the number of neurons that were initially active. Weak connections erase activity over time; strong connections amplify so much that the initial number cannot be guessed. There is a critical, medium connection strength that preserves the most information.

A similar line of reasoning could be applied to show that other information processing functions are also optimized around the critical point: the amount of information stored in the network, the network's sensitivity to inputs, and the network's computational power. We will save the details of these for chapter 4. For now, it is enough to note the first key facet of operating near the critical point: *information processing is optimized.* There is a peak in these functions, a sweet spot, and it occurs for all of them near the critical point. This is perhaps the central motivation behind studies in neuroscience that adopt the framework of criticality. The critical point seems like a natural place for a neural network to operate if its main task is processing information.

The Appearance of Emergent Phenomena

Another consequence of operating near the critical point is that the influence of one neuron extends far beyond the neurons that receive its direct connections. This expanded influence grows, or emerges, only when the network is near the critical point.

To begin to see this, let us rearrange our simple network model into a layered, feed-forward structure as shown in figure 1.4. Here distance will be the number of connections from the input layer on the left. When such a network is operating near the critical point, information about an initial number of active neurons will propagate relatively far, decaying very gradually across network layers. In contrast, for networks with connections that are too weak or too strong, information rapidly decays across layers. Thus, networks near the critical point can support the interaction of information from distant parts of the network; both local and global information can mix. In this way, activity from a single neuron can affect neurons far beyond those to which it is directly connected. This long reach of a single neuron *emerges* near the critical point and is a result of the unique way in which the neurons interact with each other there.

This expanded influence at the critical point does not merely reach through space—it also reaches through time. To illustrate, let us look at our simple model another way. What

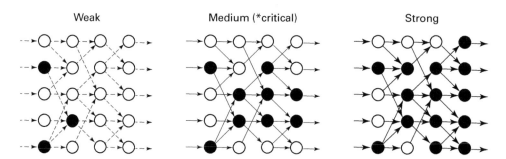

Figure 1.4
Information travels the greatest distance at the critical point. A feed-forward network with four layers is shown. Activity propagates from left to right. In each network, only two input neurons are on. When the network has weak connections, activity does not reach the output, only three layers away. Reconstruction of the input is impossible. When the network has connections of critical strength, the output layer has the same number of neurons on, allowing the input to be reconstructed. When the network has strong connections, the output is saturated, again making reconstruction of the input difficult. Each layer of connections can be thought of as distance in a brain. The ability to reconstruct the input is related to information transmitted from the input layer to the output layer, across distance.

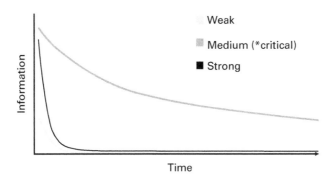

Figure 1.5
Information has the longest lifetime at the critical point. Networks with weak and strong connections cause information to be erased quickly by damping or amplifying, respectively. Networks with the critical, medium connection strengths preserve some information the longest.

is the amount of information left in the network at different times? At first, all three networks would contain the same information about their initial conditions. For reasons we explained before, this would decay rapidly in the networks with weak and strong connections. But when the connections are of the critical strength, information will decay less quickly. Remember that the network connections transmit based on probability, so they will eventually fail. This causes the information to decay, however slowly, even in the critical case. If we plot the information about the initial conditions against time for the three networks, we see that the curve with the slowest decay is produced by the network with medium strength, or critical, connections (figure 1.5). Here the activity of a single neuron echoes or reverberates much longer than the time it would take to travel through one connection. This causes long-lasting temporal correlations to *emerge* near the critical point.

In these two examples, the effect of activity in one neuron can extend in distance and time much greater than we would expect from the connections we built into the model. Ordinarily, those direct connections would only influence neurons one synapse away and one time step into the future. But when the network is poised near the critical point, these influences are enlarged. This shows how something new, not built into the model, can emerge as a result of the collective interactions among the neurons.

As we will see in the next chapter, neural network models can produce more exotic emergent phenomena like waves, pulses, and repeating activity patterns. For now, the key idea is that the critical point allows new collective patterns of activity to appear. The enhanced range and duration of information in the network facilitates mixing of activity from different parts of the network at different times. This creates novel combinations of information that would not have been possible away from the critical point.

Power Laws

Now let us look more closely at the extent to which networks at the critical point mix information from different times and from different locations. If we take the curve for information vs. time plotted in figure 1.5 and transform its axes to be logarithmic, we see that the dark gray curve from the network at the critical point now appears as a line (figure 1.6). This says that the relationship between information, I, and time, T, can be described by a

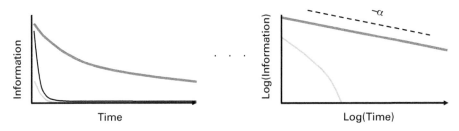

Figure 1.6
Long-tailed curves at the critical point are power laws. *Left panel*, Information vs. time, as shown in figure 1.5. *Right panel*, Plotting the log of information vs. the log of time transforms the dark gray curve into a straight line. This straight line is a power law, one of the signatures of operating near the critical point. The slope of the line, $-\alpha$, is the exponent of the power law.

simple equation like this: $I(T) = T^{-\alpha}$. This equation is a power law, and the exponent, $-\alpha$, gives the slope of the line in the double logarithmic plot. Power laws typically appear in data from systems operating at the critical point. As we will see later, even operating near the critical point can produce data that approximately follow a power law (the appendix has sections explaining power laws in more detail).

One of the most interesting facts about power laws is that they can imply some mixing occurs across the entire lifetime of the network and across its entire size. This means that even in very large networks, such as one with as many neurons as an adult's brain, it is possible for information to transmit throughout the entire network and for new information to interact with information that entered the network at its inception. Indeed, given the power law, this could remain true even if there were an infinitely large brain.

This is one of the reasons power laws are often called "scale-free." In our illustration, this means that both recent events and those from the distant past can mix; both local events and those from far away can interact. There is not one level, or scale, of the system that dominates. Rather, all scales have the possibility of influencing each other. In terms of the brain, this means that one neuron can have just as much effect on the animal's behavior as an entire cortical region. If a brain is operating at the critical point, then its dynamics really can be understood only by considering all these different levels.

A cautionary note about power laws is in order. While they are produced by systems operating at the critical point, their mere existence is not proof that a system is critical. Power laws are necessary, but not sufficient, for establishing a critical point. This is because there are several ways of producing power laws that do not rely on criticality. For this reason, establishing that a system is operating near the critical point should always involve multiple signatures from the data, and not just a power law. We will elaborate on this point in greater detail in chapter 3.

Avalanches

Information is not the only thing that follows a power law in networks at the critical point. We can also look at the cascades of activity these networks produce. If we were to initiate activity in one neuron of our model, how many other neurons would it typically activate before the cascade stopped? Figure 1.7 shows some examples. Cascades of many sizes do occur, but very small cascades occur far more often than very large ones.

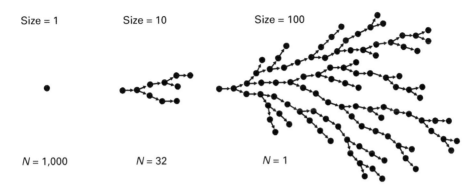

Figure 1.7
Activity cascades of different sizes. A single neuron is activated at the start, and time is unfolded from left to right. The size of a cascade is just the total number of neurons activated. N is the number of times cascades of that size are observed. In these examples, each neuron is never reactivated in a given cascade.

If we randomly activate our network to produce thousands of cascades, we can plot the number of observations of each size against cascade size as in figure 1.8. Here we can see exactly how much more frequently the small cascades occur than the large ones. If there are 1,000 cascades of size one, there will be about 32 cascades of size 10 and 1 cascade of size 100. Each time we go up in cascade size by a power of 10, we drop in cascade number by a factor of $10^{-\tau}$, or 0.0316. For reasons we will explain more in chapter 5, in the type of model we are using, this exponent τ will be −1.5 for cascade sizes. As with the previous plots for information, we could also show cascade sizes produced by a model with connections that were too weak or too strong. In both those cases, the curves would not follow the power law we see here for the critical case with connections of medium strength.

When the cascade sizes of neuronal activity are distributed according to a power law, we call them "neuronal avalanches." This term "avalanche" has a long history in physics associated with cascades in complex systems, but it unfortunately has a very different colloquial meaning. In conversation with nonscientists, the term typically evokes thoughts of huge piles of rocks or snow crashing down a mountain to bury a village. Given this, the term "neuronal avalanches" might make one think of seizures that activate every neuron in the brain. Here, though, an avalanche can be a cascade of size one and is not necessarily huge. As we saw, the small avalanches are many powers of ten more common than the extremely large avalanches.

It is worth pointing out, though, that while extremely large events are very rare with power-law distributions, they are at least possible. With the more commonly encountered Gaussian or normal distribution, it is virtually impossible to observe an event that is several powers of 10 larger than the mean. Consider human height, for example. The world average female height is about 165 cm (5 ft., 3 in.) and 95 percent of all females will fall within a range of 151–179 cm (Roser, Appel, and Ritchie 2013). A woman with a height that is 10 or 100 times 165 cm simply does not exist. Yet if such heights were power-law distributed as in our example, there would be a few women who were 16.5 meters tall and perhaps one that was 165 meters tall! Power laws tell us that such extreme events are possible, even if rare.

If this model applies to the brain, we would expect that activations of extremely large numbers of neurons would occur, though very rarely. Interestingly, about 1 percent of people

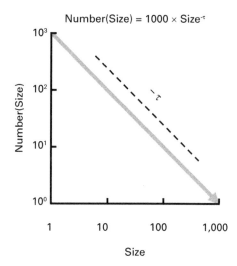

Figure 1.8
A power law for avalanche sizes. The model is activated at randomly chosen neurons over 24,000 times. The y-axis is the number of times a cascade is observed; the x-axis is the size of the cascade. When the network is critical, the distribution of cascade sizes will follow a power law whose exponent τ will be -1.5. Note that here both axes are logarithmic, with tick marks jumping by powers of ten.

in the United States will experience a seizure at some time in their lives (Zack and Kobau 2017), and this is typical of global statistics. The possible connections between operating near the critical point and seizures is a growing area of research that we will revisit in chapter 6.

A Phase Transition

In addition to being connected to power laws, the critical point is also related to phase transitions. If we return to our plot of information versus connection strength (figure 1.3), recall that to the left of the dashed line were model networks with weak connections and to the right were those with strong connections. These differences in connection strengths lead to differences in activity. A network with weak connections will have little to no activity, even if it is being driven occasionally from some outside source. This is the inactive phase. On the other hand, a network with strong connections will be nearly continuously active, even if it is driven rarely; this is the active phase. The peaked information curve that we saw in figure 1.3 becomes even sharper for larger networks; we will explain why in chapter 4. Thus, poised right between these phases is a narrow region where the critical point resides. This is the phase transition, at the dividing point between the inactive phase and the active phase (figure 1.9).

While the model we have been discussing here has a transition between an inactive phase and an active phase, this is sometimes described as a transition between disorder and order (Brush 1967; Fraiman et al. 2009; Marinazzo et al. 2014). Other types of transitions are possible—many models have explored the transition between an asynchronous oscillating phase and a synchronous oscillating phase (Kuramoto 1984; Kitzbichler et al. 2009; Breakspear, Heitmann, and Dafferts 2010; Dalla Porta and Copelli 2019). There is

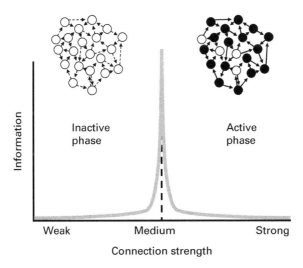

Figure 1.9
The critical point marks a phase transition. To the left, with weak connection strengths, the network is largely inactive. To the right, with strong connection strengths, the network is nearly always active. The critical point lies at the transition between these phases. The phase transition is marked by the sharply peaked information curve. As networks become larger, the information curve becomes sharper as shown here.

much discussion about what type of model could best capture the observed patterns in neuronal data, and this is likely to be at the frontier of research for years to come.

From a Model to Data

So far, we have just been talking about a very simple model to illustrate the critical point and its implications. As we will see, there are several types of neural models that have been developed to describe phase transitions. All of them produce power-law distributions and show sharply peaked functions—hallmarks of criticality.

Is there any reason to believe that these concepts have relevance for real brains? In fact, researchers have used the framework of the critical point to interpret their data from a variety of species, including worms (Aguilera, Alquézar, and Izquierdo 2017), zebrafish (Ponce-Alvarez et al. 2018), salamanders (Tkačik, Mora et al. 2015), turtles (Shew et al. 2015), mice (Scott et al. 2014), rats (Beggs and Plenz 2003; Friedman et al. 2012), monkeys (Petermann et al. 2009), and humans (Haimovici et al. 2013; Priesemann, Valderrama et al. 2013; Shriki et al. 2013). These data often have distributions that follow power laws, as well as other signatures of operating near the critical point. Before we discuss some of these studies in later chapters, it will be helpful to describe how some of the earliest research began to show that the concept of criticality would be applicable to the brain.

Evidence of a Critical Point

By the early 2000s, several labs sought to explicitly test the predictions of critical models for the brain. We will now explain three of these first tests. Recall that in our simplified neural network model, activating a single neuron left a trace of information that died out very gradually with a power-law decay (figures 1.5 and 1.6). This slow decay suggested a way to test if living neural networks operated near the critical point. For example, during magnetoencephalographic (MEG) recordings of a human subject there are often spontaneous fluctuations in the signal. These fluctuations might be viewed as perturbations that could reveal the decay times of the brain. Typically, a large increase in, say, the alpha power band (8–12 Hz) will persist for a while before disappearing. How slowly does such a fluctuation die out?

In 2001, Klaus Linkenkaer-Hansen and colleagues (Linkenkaer-Hansen et al. 2001) examined these types of decays. They reported that they indeed followed a power law (figure 1.10) and that the exponent for the power law was the same across subjects. This suggested a general principle governing the temporal correlations, and they explained their results within the framework of criticality. To ensure that this result was not an artifact, they examined temporally shuffled data and showed that they had an absence of temporal correlations. This was an important paper, demonstrating that data from humans were consistent with predictions that the brain operated near a critical point. (The appendix has a section explaining in more detail how long-range temporal correlations are measured.)

At nearly the same time, Greg Worrell and colleagues were recording electrical signals from the hippocampi of human epilepsy patients (Worrell et al. 2002). Between seizures, for five hours at a time, they looked at the amplitude of the signals produced at a single electrode. The signals were intermittent, with many small-amplitude events and very few large-amplitude ones. They found that the amplitude squared, a measure of the energy, followed a power-law distribution (figure 1.11B). In addition, they showed that the measured exponents could not have been explained by a random walk process.[2]

While their data were consistent with the concept of criticality, each patient had their own exponent. That meant that a single unifying equation governing the patients did not apply. Worrell and colleagues noted that many of the different critical models in the literature at that time had exponents that varied, depending on the exact parameters of the system (Worrell et al. 2002). They argued that their seven epilepsy patients could have different

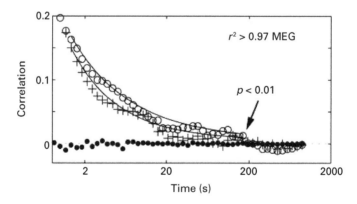

Figure 1.10
Power-law decay from human brain recordings. Alpha oscillations were recorded via MEG and their autocorrelation was plotted against time. Solid lines give power-law fits to eyes open (circles) and eyes closed (crosses) conditions averaged across 10 subjects. Note that the x-axis is logarithmic; the y-axis is not, which causes power laws to appear as curves here. Black dots show shuffled data, effectively at zero. The significance of the correlation compared with zero at a time lag of 200 s is shown. Adapted from Linkenkaer-Hansen et al. (2001).

A

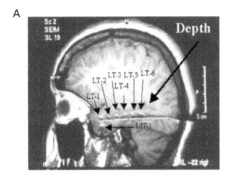

B

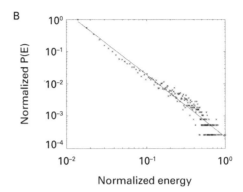

Normalized energy

Figure 1.11
Power-law distribution of energy fluctuations in human hippocampus. *A*, MRI image showing placement of a six-contact depth electrode within the left temporal lobe. Arrows with the labels LT1, LT2, etc., indicate the six contacts, each separated by 1.0 cm. *B*, The normalized probability density of energy fluctuations occurring in the period between seizures, plotted in log-log coordinates. The straight line has a slope of −1.9. Adapted from Worrell et al. (2002).

conditions, so it might not be surprising that the exponents varied. Despite the caveat, this work again suggested that human neurological data had a signature of criticality.

Both studies just mentioned relied on data from a single recording channel measured over a long duration. While these data allowed measurements of temporal correlations and even the sizes of events, they did not describe how activity propagated in space. For that, multichannel data were needed.

In 2003, Dietmar Plenz and I analyzed data from rat cortical cultures and slices placed on arrays with 60 electrodes arranged in a square grid (figure 1.12A), allowing for spatial analyses (Beggs and Plenz 2003). These cortical tissues were spontaneously active, and in the course of 10-hour recordings a given electrode detected up to several hundred thousand negative local field potentials (nLFPs) (figure 1.12B). Each nLFP reflected the synchronous activity of many neurons firing spikes near the electrode (Petermann et al. 2009). To detect activity propagation across space, we binned the data into 4-ms segments; this matched the average time it took for a signal to travel from one electrode to its nearest neighbor. When activity across the entire array was viewed at this temporal resolution, it was apparent that the nLFPs did not occur in isolation but rather in spatial and temporal groups that propagated across the array (figure 1.13C). With this framework, we defined an avalanche as a sequence of consecutively active time bins, bracketed at the start and end by inactive bins (figure 1.13B). The avalanche size was the total number of electrodes with an nLFP signal that crossed a threshold, and the avalanche duration was the total number of time bins from start to end.

Avalanche sizes in all data sets were distributed according to a power law with an exponent of −1.5 (figure 1.13D). In addition, the initial portion of the avalanche duration distribution approximately followed a power law with an exponent of −2 (figure 1.13E). This was of interest because these were exactly the exponents predicted by Eurich, Herrmann, and Ernst (2002) in an avalanche neural network model. These were also the exponents of a critical branching process, where one active neuron was expected to lead to one other active neuron in the next time step. To probe this connection, we directly measured the

A   B

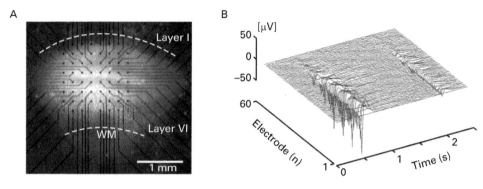

Figure 1.12
Spatial and temporal activity from a cortical slice network. *A*, A cortical slice culture placed on a 60-electrode array. Electrodes appear as small black dots at the end of straight wires; upper and lower cortical layers are marked by dashed lines. *B*, Example bursts of local field potentials occur near 0.5 and 2.0 seconds. The first burst has a larger negative amplitude. Both bursts involve the majority, but not all, electrodes. Negative local field potentials that exceeded a threshold of −3 standard deviations were marked as events that contributed to avalanches. Adapted from Beggs and Plenz (2003).

average number of descendants from a single active electrode, called the branching ratio, σ, and found that it was indeed close to 1, as expected for a critical branching process (figure 1.14A).

Several controls supported the soundness of this approach. First, shuffled data did not produce power-law distributions, showing that they were not the result of chance. Second, the power-law distributions were unchanged in shape if every other electrode was deleted from the analysis, or if half the electrodes were deleted from the analysis. This indicated that the results were scale invariant, as would be expected for a system operating near the critical point. Third, all the data sets, including cortical slice cultures and acute cortical slices, showed approximately the same exponents. Fourth, the power law for avalanche sizes had the same form if avalanche size was measured by the number of nLFPs observed or by the amplitude of the nLFPs (figure 1.13D). Fifth, the power law was disrupted when picrotoxin was applied, which blocked inhibition, and it recovered after picrotoxin was washed out. This showed that the power laws were dependent on the proper balance between excitation and inhibition and were not a default condition unrelated to a critical point. This also demonstrated that the network could be tuned away from the critical point to a different phase. Collectively, these controls strongly suggested that the data were consistent with a phase transition produced by a critical branching process.

In this same paper, we also connected a critical branching process with optimal information transmission. Using a simple feed-forward neural network model (figure 1.14B), we demonstrated that mutual information between the input and the output of the network was maximized when the branching ratio was at the critical value of 1 (figure 1.14C). Note that this feed-forward neural network model is very similar to the one discussed previously (figure 1.4). It is also consistent with the earlier work of Greenfield and Lecar (2001), who used a slightly different computational model but came to the same conclusions.

Despite the quality of the data, the controls and the connection to optimal information transmission, there were several shortcomings of this study. The nLFP recordings did not reveal what was happening at the single neuron level; single spike recordings would be

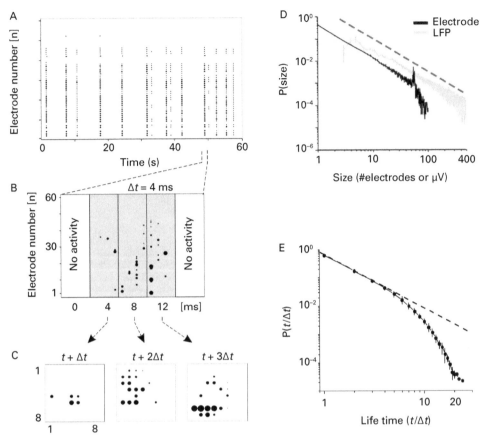

Figure 1.13
Neuronal avalanches produce power-law distributions. *A*, A raster plot of local field potential activity on the array over time. Electrode number is along the y-axis, time is on the x-axis. Apparently synchronous bursts of negative local field potentials appear as columns of dots. *B*, Each burst is composed of activity spanning several time steps. An avalanche is defined as activity in consecutive time bins ($\Delta t = 4$ ms), bracketed by no activity at the beginning and end. Diameter of dots is proportional to amplitude of negative local field potentials. *C*, Activity is replotted on the square electrode array, showing spatial locations. *D*, Neuronal avalanche sizes are distributed as a power law, whether size is measured by the total number of electrodes exceeding threshold or by the total amplitude of the local field potentials in the avalanche. The dashed line shows a slope of −1.5. *E*, Neuronal avalanche lifetimes are distributed as a power law with an exponential cutoff. The dashed line shows a slope of −2. Adapted from Beggs and Plenz (2003).

necessary for that. While 60 channels were more than had been used for studies of this sort at that time, even more channels would have allowed finer examination of the activity in the tissue. Undoubtedly, many neurons were missed and issues about subsampling the data would later be raised (Levina and Priesemann 2017). Lastly, the data were not collected from an intact animal, and there obviously was no behavior, so questions remained over the general validity of the conclusions. Would these findings apply to an animal in a natural setting? Or was the phenomenon of the critical point a curiosity limited to preparations found only in a lab? All of these are valid points. Much work remained to be done to assess whether the concept of criticality could be established with greater rigor, or if it could find broader application within neuroscience.

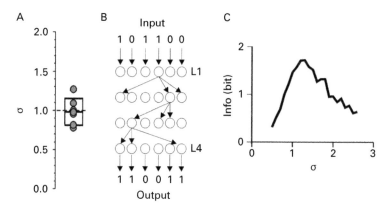

Figure 1.14
The measured branching ratio σ suggested optimal information transmission near the critical point. *A*, The branching ratio measured for seven slice networks had an average value of σ = 1.04 ± 0.19, near the critical value of 1.0. *B*, A simple feed-forward model network with four layers (L1 to L4) was used to test the impact of the branching ratio on information transmission. For clarity, not all connections are shown. *C*, Information transmission showed a peak slightly above a branching ratio of 1. As network size was increased, this peak moved closer to the critical branching ratio. Adapted from Beggs and Plenz (2003).

Nevertheless, these early studies did manage to excite interest in some quarters by hinting at a link between physics and the brain. The statistical physics that began in the 1800s developed into an extremely successful theory that could describe the collective properties of an ensemble of *particles*, like gas molecules. A single equation could relate their pressure, volume, and temperature. In addition, a detailed knowledge of how these properties changed at the critical point was established. But could a similar theory be developed to describe the collective properties of an ensemble of *neurons*, not in terms of their physical characteristics like pressure and volume but in terms of their information? Could we have a physics of neuronal ensembles that could tell us how much information they stored or transmitted, or their computational power or sensitivity to inputs (Beggs 2015)? And could we relate these quantities to the critical point? This remained to be seen. But at least now it was possible to ask these questions and begin to think about the experiments that might answer them.

The Criticality Hypothesis

Now that we have seen some of the implications of operating near the critical point and early experimental data, I can more fully describe the criticality hypothesis. This will be useful for future discussions. I take the criticality hypothesis to mean this: *When a network of neurons collectively interacts near a critical phase transition point, it simultaneously optimizes multiple information processing functions.* In chapter 4, we will explore these information processing functions in more detail; in chapter 8, we will discuss what is meant by being near the critical point, and in chapter 9, we will describe why this hypothesis is most applicable to the mammalian cortex.

The two main consequences of being near the critical point are *scale-free properties* and *universality*. As we just saw, scale-free properties are hypothesized to lead to optimal information transmission. And as we will soon see in chapter 4, they are also thought to optimize dynamic range, sensitivity to inputs, information storage, and computational power. For a finite brain, these benefits can occur even if it is poised near the critical point, but not exactly at it. In fact, the finite size of the brain makes it easier to enjoy some of these benefits—just being near the critical point will produce optimality over the scale of the brain, or a cortical region, without having to reach to infinity. Reaching to infinity would require being exactly at the critical point.

The other consequence, universality, will be discussed in detail in chapter 5. For now, let us just say that whenever the same phenomenon, like neuronal avalanches, occurs across many different scales—the scale of a network of hundreds of neurons, the scale of a cortical region, or the scale of the entire cortical mantle—then the phenomenon is likely to be captured by a simple model. If the model were detailed and dependent on the particulars of the neural tissue at each scale, it likely would not show the same behavior across scales. Thus, the scale-free properties of neuronal avalanches strongly suggest that some relatively simple model will be enough to capture the main features of their behavior. Just as such a model will be independent of scale, it is also expected to be *universal* and independent of other details like the species, the type of neurotransmitters, and the cell types. Universality is not necessarily a benefit to the brain itself, but it is a fortunate consequence for those of us trying to understand how brains work. And gladly, such nearly universal behavior is expected to be present even when a brain is close to, but not exactly at, the critical point.

Being close to the critical point is therefore enough to produce scale-free properties and universality over a finite range. But not being exactly at the critical point also means that theories from physics that were developed to deal with the critical point will not strictly apply to the brain. What is needed is a new principle to describe how the benefits of the critical point are modified under the conditions of finite size and constantly changing inputs that real brains experience. In chapter 7, we will explain a candidate for this principle, quasi-criticality, in more detail, describing how some of its predictions have already been experimentally tested.

Objections and Responses to the Criticality Hypothesis

The criticality hypothesis has generated much discussion and criticism. Let us now briefly turn to some of the more common questions about this research—remarks I have heard often from grant reviewers and paper referees!

One might object that if signatures of operating near a critical point show up everywhere, in so many species and with so many methods, then surely this finding is too general to be interesting. Isn't it rather something trivial like the observation that all household objects have mass? That is, so what?

To put this view in perspective, let's consider wings that appear in nature. They are found in seeds from maple trees and on dragonflies, flying fish, pterodactyls, Draco lizards, birds, and bats. Even flying squirrels and some snakes can be said to have wings of a sort. Yet this abundance of wings does not necessarily make them uninteresting. Instead, it suggests a general operating principle that is exploited by different species under diverse circumstances. Does this general principle make the study of wings, how they work, why they differ, and how they can be used in technology an uninteresting activity? I don't think so. In the same way, understanding why the brains of so many species seem to operate near a critical point may reveal common principles used by nature for processing information. Understanding these principles may help us in fixing damaged brains or in designing artificial neural networks or more effective computational devices (Stieg et al. 2012; Cramer et al. 2020).

One could also object that the brain should naturally avoid the extremes of being overly damped or excessively amplified. Of course, it must be somewhere in the middle, so the critical point is merely a broad and uninteresting default setting. Where else would we expect it to be?

Actually, the critical regime is extremely narrow and not what we would expect by chance. First, when data are randomly shuffled, they no longer produce power-law distributions characteristic of the critical point (black dots in figure 1.10). Neuron spike times randomly jittered by as little as 20 milliseconds also disrupt signatures of operating near the critical point (Friedman et al. 2012). Second, slight perturbations by pharmacological manipulations (Beggs and Plenz 2003; Ponce-Alvarez et al. 2018) or changes in stimulus intensity (Shew et al. 2015) can push a network away from near the critical point. This sensitivity suggests that it would not be easy to maintain operation near the critical point. As we will see later, there are active processes to homeostatically regulate neural networks near the critical point. Third, operation near the critical point is only seen in select brain structures in specific circumstances: when visual stimulation is delivered to turtle cortex, visual areas are close to critical, but nonvisual areas are not (Shew et al. 2015). All of this argues against the idea that operating in the vicinity of the critical point is merely a broad default state into which every brain region falls. This naturally prompts interesting questions. Is being near the critical point important for information processing? How close must a network be to the critical point to effectively reap the benefits in information processing? How often do actual networks fall out of this narrow range? Why do some brain regions and not others hover near a critical point? Answering these is not trivial.

It is often stated that neuroscience research into the critical point is controversial. If the evidence is so compelling for this picture, then why isn't there complete consensus? Why does there continue to be a group of scientists voicing strong concerns?

It is the nature of science to be skeptical about truth claims and to submit them to rigorous tests. The claim that the cortex operates near a critical point is a sweeping one, encompassing optimum information processing, neurological health, and near universal application across species. Extraordinary claims require extraordinary evidence. Some of the earliest work on criticality relied too heavily on the mere existence of power laws. Skeptics rightly noted that power laws, by themselves, do not prove criticality (Mitzenmacher 2004). For example, successive fragmentation is one of several processes, including combinations of exponential curves (Reed and Hughes 2002) that can produce power laws yet do not indicate criticality (Beggs and Timme 2012).

In response, the field developed more sophisticated methods for quantifying criticality that did not rely solely on power laws (Friedman et al. 2012; Wilting and Priesemann

2018). Some of these new methods were acknowledged by critics as true indicators of the critical point (Touboul and Destexhe 2017; but now see Destexhe and Touboul 2021). There were also questions about whether or not the neuronal data were indeed following power laws (Clauset, Shalizi, and Newman 2009). A very close examination of the results indicated that many data sets in fact were (Yu et al. 2014).

Perhaps because of these responses, current skepticism no longer centers around power laws but has moved on to other topics, like whether operating near the critical point is behaviorally relevant (Fagerholm, Lorenz et al. 2015) or whether the cortex operates exactly at the critical point or slightly below it (Wilting and Priesemann 2019a). This suggests the first waves of criticism were answered by better methods and experiments. Research is now addressing these subsequent questions in what appears to be fruitful scientific dialogue.

It is desirable that skeptics continue to raise questions in these areas if we are to refine our picture of how the brain works. The history of science shows that many successful theories (Copernicus's heliocentrism, Einstein's general relativity [Sherwood 2011], Wegener's continental drift [Hallam 1975]) experienced periods of controversy, sometimes intense, before being widely accepted. The existence of controversy, by itself, says very little about whether an idea is good or bad. Rather, it is how these controversies are handled that indicates whether science is progressing or not.

Chapter Summary

A simple model shows that with the proper connection strengths, neurons can collectively operate near a critical point. Near this narrow point, activity is poised between a phase where activity is damped and a phase where it is amplified. At this phase transition, the simple model attains optimal information processing and emergent dynamics. One signature of operating near the critical point is a distribution of avalanche sizes that approximately follows a power law, as seen in data. While the hypothesis that the cortex operates near a critical point has generated some skeptical questions, the field has responded with new methods and data. The promise of a unifying framework for collective neural activity continues to generate growing research interest.

2

Emergent Phenomena

One of my favorite times in the academic year occurs in early spring when I give my class of extremely bright graduate students, who have mastered quantum mechanics but are otherwise unsuspecting and innocent, a take-home exam in which they are asked to deduce superfluidity from first principles. There is no doubt a special place in hell being reserved for me at this very moment for this mean trick, for the task is impossible. Superfluidity . . . is an emergent phenomenon—a low-energy collective effect of huge numbers of particles that cannot be deduced from the microscopic equations of motion in a rigorous way and that disappears completely when the system is taken apart.

—Robert Laughlin, Nobel Lecture: "Fractional Quantization"

It is mesmerizing to watch a flock of starlings, called a murmuration, move through the sky before dusk (figure 2.1; see here for a video: https://petapixel.com/2017/12/13/eye-popping -starling-murmuration-captured-black-white/). The murmuration gradually changes shape, with tendrils budding off and then rejoining. Within the main body, there appear to be undulations rippling swiftly across its length. How can such a large entity seem to have a life of its own when it is made up of thousands of independent birds?

Research tells us that each bird's heading is determined by a simple averaging of the directions of its seven nearest neighbors (Ballerini et al. 2008; Young et al. 2013). Simulations using this rule produce flocks that behave in the same way as actual flocks (Hemelrijk, van Zuidam, and Hildenbrandt 2015). Apparently, the complex patterns we see in the flock at a large scale are merely a consequence of a simple rule that operates only at a small scale. We can say that these large-scale properties emerge out of the interactions between starlings. And this emergence is somewhat mysterious; even with knowledge of the simple rule, it is extremely difficult to predict what these large-scale properties will be without running the simulation.

Another interesting feature of these flocks is that they are operating near a critical point. This conclusion is supported by the observation that the distance at which different starlings' headings are correlated, called the correlation length, is as large as the flock itself (Cavagna et al. 2010; Bialek et al. 2014). This is like what we saw in the simple neural network model from chapter 1: the distance at which information could travel through the network became very large at the critical point.

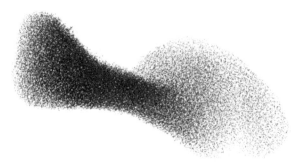

Figure 2.1
A murmuration of starlings. At this scale, each bird appears as a dot. Photograph adapted from: https://www
.bbc.com/news/science-environment-29599792.

In this chapter, I will argue that one of the best ways to grasp the interesting world of emergent phenomena is by first understanding the critical point. Comprehending emergent phenomena can be difficult because we have been trained to analyze most systems, including complex ones, by taking them apart. The problem is that most emergent phenomena disappear when they are taken apart. How then can we understand them? We must learn to shift our focus away from the parts themselves and onto the interactions between the parts. Sometimes these interactions do not even depend on the types of parts that are present. In describing emergent phenomena, we will have to address some issues that are essentially philosophical in nature. Rather than diving into technical details, I will try to present these topics from an intuitive point of view. This coverage is not meant to be exhaustive, as there is a vast philosophical literature on emergence (O'Connor 1994; Favela 2019). The discussion that follows is intended as an introduction to some of the issues surrounding emergent phenomena that are relevant to the critical point and the brain.

Methodological Reductionism

When faced with a system as complex as the brain, how can one begin to understand how it works? A very natural approach is to break it down into its components. Dissect it, catalogue the parts, note the connections between the components and then start manipulating them, one at a time, to see how they affect the system as a whole. This approach is called methodological reductionism, and it has been enormously successful in neuroscience.

One of the reasons for this success is that the manipulations reductionism prescribes can lead directly to a description of causality in the system. For example, in trying to understand the action potential, Hodgkin and Huxley (1952) applied tetrodotoxin to block the flow of sodium ions. They could also prepare solutions devoid of sodium. Without sodium, there was no inward current to cause the depolarizing portion of the action potential; sodium therefore was the cause of the inward current. Their manipulations led to a clear link between an ion and the thing that it caused.

In exciting experiments aimed at probing the causes of decisions, neuroscientists have found that injecting current to affect a single neuron can significantly bias the outcome of a behavior. For example, when leeches move, they either swim or crawl. Which behavior is exhibited depends largely on the activity of a single cell in a ganglion. When this neuron is depolarized, the leech is more likely to crawl; when it is hyperpolarized the leech is more likely to swim (Briggman, Abarbanel, and Kristan 2005). In a similar manner, when monkeys are faced with a binary perceptual decision that has approximately equal odds of occurring, microstimulation of a neuron encoding the percept can significantly bias the decision (Salzman, Britten, and Newsome 1990). Both these experiments show that our understanding of complex behaviors can be enhanced by causal manipulations at the single neuron level. Activity in these neurons can cause certain behaviors.

Technology often drives methodological reductionism; if we have the means to affect a single ion, a single neuron or a single gene, then we can explore with relative ease the consequences of manipulating those things. And if most of our experiments are designed around the available technology, our experiments could create the impression that it is always the lower-level factors that are the most fundamental. Of course, we know that a brain is composed of many parts, from transmitters at the molecular level to synapses, circuits, and cortical regions, and we may have some tacit acknowledgement that all these levels interact with each other. But if most of our experiments involve causal manipulations at the lowest levels, we could gradually develop the view that the arrow of causality only runs from the bottom up. Could higher levels ever cause things to happen at the lower levels? Let's explore this by thinking about a very simple emergent phenomenon.

The Wave as an Emergent Phenomenon

One of the most commonly observed emergent phenomena occurs in packed football or soccer stadiums and is called "the Mexican wave," or the "stadium wave." In this wave, a narrow column of spectators, from near the field to the highest seats, rise from being seated, stand to full height, raise their arms, let out a cheer and then sit down (figure 2.2). This is followed by an adjacent column of spectators doing the same. Each column seems to trigger another, giving rise to a wave of cheers traveling around the stadium. Videos of the wave are easily found on YouTube (for a typical wave, see here: https://www.youtube

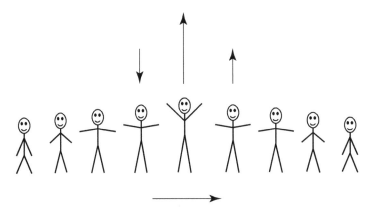

Figure 2.2
The stadium wave. The cartoon shows one row of people in a stadium wave propagating from left to right. The peak of the wave separates those who are rising on the right from those falling on the left.

.com/watch?v=Ggl0qGJ2zCg; for slow and fast waves, see here: https://www.youtube .com/watch?v=FTlnLyeSUMw). Analyzing data from several videos, Farkas, Helbing, and Vicsek (2002) found that the wave had a width of about 15 seats and traveled at about 20 seats per second. Interestingly, it usually travels clockwise when viewed from above.

The wave is clearly an emergent phenomenon because it appears, or emerges, only when large numbers of people interact. Although this experiment has not been done to my knowledge, if you blindfolded the people and gave them earplugs, it would almost certainly destroy the wave. The wave also has properties that are different from the constituent parts that make it up. For example, the wave is larger than a single person and faster than a single person. The wave also changes direction, or reflects, when it hits a boundary like a gap in seating. Individual people do not reflect at such boundaries. If the wave is representative of emergent phenomena, then it seems that these phenomena depend on interactions among smaller-scale components and that new properties appear at the larger scale, different from those at the smaller scale.

Let's now think about what knowledge of the constituent parts reveals about the emergent phenomenon. Imagine if we took a single person from the stadium, brought them into a lab and asked them to sit down, stand up and cheer, and then sit down again, hundreds of times. How much would this tell us about the wave? It is not clear that this would let us deduce the wave's size, speed, and reflective properties. Deeper study of a single person, or even many people as individuals, probably would not help us much either. To properly study the wave, we would have to see thousands of people as they interacted; we would have to catch the wave in its natural environment. Simulating a stadium full of interacting people could get us further and reveal large-scale properties that we might not have expected. As Philip Anderson noted in a famous article, "More is Different" (Anderson 1972):

The behavior of large and complex aggregates of elementary particles, it turns out, is not to be understood in terms of a simple extrapolation of the properties of a few particles. Instead, at each level of complexity entirely new properties appear, and the understanding of the new behaviors requires research which I think is as fundamental in its nature as any other.

From this, it appears that the properties of emergent phenomena are not easily deduced from knowledge of the parts that make them up.[1]

But the parts must matter somehow. What if we took English people at a soccer game and replaced them with Brazilians? The wave probably would remain basically the same. In a strange thought experiment, it seems possible that a stadium of well-trained dogs might be able to sustain a wave. Let us consider one more change, to the seats. Imagine we switched from a rectangular grid of seats to a triangular grid of seats, where the seats in the next row are lined up at the gaps in the previous row. These examples serve to illustrate that the properties of emergent phenomena do not depend on some of the smaller-scale details. They are effectively cut off from some causal effects at lower scales. Note that this is not just saying that the causal arrow typically revealed by methodological reductionism is incomplete. This is saying that with emergent phenomena there will be times when this causal arrow is actually wrong. Lower scales do not always drive the higher scales (figure 2.3).

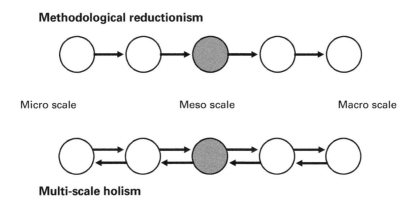

Figure 2.3
Reductionism vs. holism. Each circle represents a scale in the system, ranging through micro, meso to macro. The circles could be, for example, transmitter molecules, neurons, microcircuits (gray), cortical regions, and mind. Methodological reductionism assumes that only lower levels can produce causal effects in higher levels. Multi-scale holism assumes that causality can extend from upper levels, where emergent phenomena arise, downward to the micro level. In this view, understanding must include all scales, because all scales interact.

This arrow of causality deserves further consideration. Let us think about how the wave, a relatively large-scale structure, interacts with things at smaller scales. As it moves throughout a stadium, could the wave cause a person to stand up and spill her beer? If this is possible, and it seems to make sense to think about it this way, then we would have an entity at a larger scale causing effects at a lower scale. In our previous discussion we noted that this is not the direction of causation typically revealed by methodological reductionism. To distinguish this surprising type of effect, philosophers have introduced the term "downward causation" (Bedau 2002; Bishop 2008), and it has generated much discussion. While this concept is not agreed upon by all philosophers and scientists (Andersen et al. 2000), the fact that it is being considered fervently marks an interesting sea change from the days when reductionism was the dominant view. A substantial number of scientists and philosophers now believe that emergent phenomena can causally influence things at scales smaller than themselves.

Finally, the wave has no mass by itself—the total weight of the stadium at 2:05 PM when there is no wave is the same as at 2:06 PM when there is a wave. Yet individual people clearly have mass. This means an emergent phenomenon, merely through its order and arrangement, can causally influence material objects. This type of view hearkens back to the work of Ludwig Boltzmann (Sharp and Matschinsky, 2015), who showed that the configuration of particles, as quantified by their entropy, had to be considered just as much as the mass and the energy of a system. While this may sound strange, it is not unique to the wave and may hint at a way to think about the mind-body problem. To illustrate, would it be possible for a person to make up their mind to be more optimistic, thereby becoming less depressed and eventually changing the levels of serotonin in their brain? Emergent phenomena, though apparently immaterial, can by their arrangement exert effects on material things.

Given our examination of the wave, let us try to summarize the features of emergent phenomena. These features will guide our continued discussion.

Emergent phenomena:

1. Arise at a scale larger than their smaller components

2. Depend on the interactions of their smaller components

3. Have properties that differ from those of their smaller components

4. Have properties that are not easily deduced from knowledge of their smaller components

5. Are independent of many features of their smaller components

6. May causally influence their smaller components (downward causation)

7. Are without mass, yet through their arrangement can influence things with mass

Emergent Phenomena in the Brain

The stadium wave is interesting, but do such things occur in the brain? In fact, they do (figure 2.4). Waves occur in visual areas when awake monkeys are presented with a visual stimulus (Muller et al. 2014; Fries 2009; Davis et al. 2020). Waves of oscillations appear in monkey primary motor cortex immediately preceding movement (Rubino, Robbins, and Hatsopoulos 2006). In all cases, waves are thought to convey information that is relevant for behavior (Muller et al. 2018). During development, waves occur spontaneously in the retina (Wong 1999) and across cortex (Garaschuk et al. 2000). Here, waves are thought to play a role instructing new synaptic connections to form (Adelsberger, Garaschuk, and Konnerth 2005; Firth, Wang, and Feller 2005). During locomotion, activation of both the hippocampus and entorhinal cortex appears as a traveling wave (Lubenov and Siapas 2009; Hernández-Pérez, Cooper, and Newman 2020). This wave is thought to help track one's location and support memory formation (Dickson et al. 2000; Blair, Welday, and Zhang 2007; Burgess, Barry, and O'Keefe 2007). A variation of the typical straight wave is the spiral wave, spinning around a point (figure 2.5); these have been reported in cortex under anesthesia and during sleep (Huang et al. 2010). Interestingly, the types of emergent phenomena that appear can be tuned by changing the balance of excitation to inhibition and by introducing neuromodulators like acetylcholine (Huang et al. 2004).

In addition to waves, there are pulse-like patterns that travel in cortical slices (figure 2.6). These differ from waves because they do not always extend across the width of the cortical area in which they move. They can be more compact than waves and are seen to annihilate when two of them collide (Wu, Guan, and Tsau 1999). It is not yet known if such pulses travel in the intact cortex.

There are simpler forms of emergent phenomena in the brain as well. The most basic would be synchrony; it occurs when a group of neurons all fire action potentials at roughly the same time. Synchrony is thought to play a role in grouping together neurons in time that represent different features of the same object (Singer 1999), and it has been widely observed. By firing synchronously, the information carried by a group of neurons may be more likely to depolarize and fire downstream neurons (Fries 2005).

While a synchronous firing of neurons might occur only once, oscillations require multiple synchronous firings, often at a characteristic frequency. Oscillations may function like synchrony but over longer windows of time and over greater distances where axonal

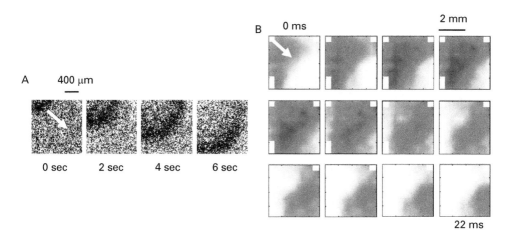

Figure 2.4
Waves in retina and cortex. *A*, A wave in the retina propagates from the upper left to the lower right, as shown by the white arrow. This is a relatively slow wave, taking 6 seconds to move about 1 mm. Images were taken from a postnatal day 0 rabbit retina; the dark region indicates the wave front where Ca^{++} concentration is elevated. Image adapted from Zhou and Zhao (2000). *B*, A wave in primary motor cortex from an awake, behaving monkey propagates from upper left to lower right as shown by the arrow. Local field potential peaks, recorded from an electrode array, are represented by darker shades, while the troughs are lighter. Each frame gives the position on the array every 2 ms. This wave travelled about 6 mm in 22 ms. Image adapted from Rubino, Robbins, and Hatsopoulos (2006).

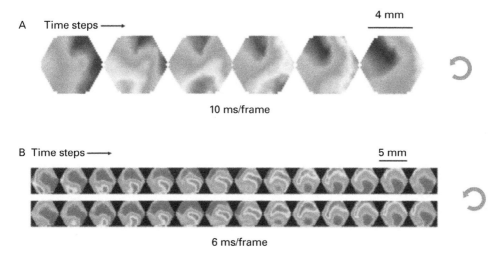

Figure 2.5
Spirals in cortex. *A*, A spiral wave recorded from rat cortex in vivo after bicuculline and carbachol were applied. Spirals also occurred spontaneously during anesthesia. Voltage-sensitive dyes were used to visualize regions of higher voltage (lighter gray) with a 464 optical detector array placed above the pial surface. A wave is seen moving counterclockwise, from lower left to upper right. Image adapted from Huang et al. (2010). *B*, A spiral from a tangential slice of rat cortex. The wave begins at the bottom of the hexagonal window and rotates counterclockwise. Frames start from upper left and go to lower right. Images recorded using the same apparatus as in *A*; bathing solution contained bicuculline and carbachol. Image adapted from Wu, Huang, and Zhang (2008).

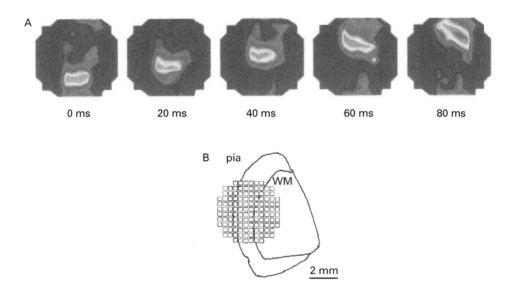

Figure 2.6
A traveling pulse in cortex. *A*, A pulse-like wave travels from bottom to top in a coronal slice of rat cortex. Higher voltages are given by lighter shades; each frame is 20 ms. *B*, The recording set up, showing position of 124 photo-diode array, white matter (WM) and pial surface of cortex (pia) of the slice. Recordings were made with voltage sensitive dyes. Pulses could be observed in normal cerebrospinal fluid. Modified from Wu, Guan, and Tsau (1999).

conduction delays could cause difficulties in producing a one-time synchronous event (Fries 2005; Buzsaki 2006). By oscillating at a characteristic frequency, many distant neurons can be made to fire in synchrony repeatedly for relatively long intervals. There are several noted frequency bands, each thought to play a different role in coordinating activity. For example, oscillations in the gamma band (~30–80 Hz) occur in the visual system and have been proposed to bind together different features of the same object, thereby representing a single percept (Eckhorn et al. 1988; Gray et al. 1989). Oscillations in the theta band (~8–12 Hz) occur throughout the hippocampus and have been shown to play a role in encoding the animal's position as it approaches and passes through a given location (Buzsáki 2002; Buzsáki and Draguhn 2004).

The final emergent phenomenon in the brain we will consider here are repeating patterns of spike or local field potential activity. These consist of a group of neurons firing in a sequence, sometimes with precisely timed intervals between the spikes. For example, we could observe neuron A firing first, followed 5 milliseconds later by neuron B firing, then followed 7 milliseconds later by neuron C firing. If such a sequence were to occur significantly more often than expected by chance, it would be identified as a repeating pattern. These have been reported to occur when birds sing well-rehearsed portions of a song (Hahnloser, Kozhevnikov, and Fee 2002) (figure 2.7), when rats travel through the same portion of a maze (Louie and Wilson 2001), and they replay at higher speed during sleep when rats (Wilson and McNaughton 1994) or birds (Dave and Margoliash 2000) are dreaming about the previous day's activity. Although there has been some controversy over the proper way to identify statistically significant repeating patterns, they have been widely reported under a variety of circumstances.

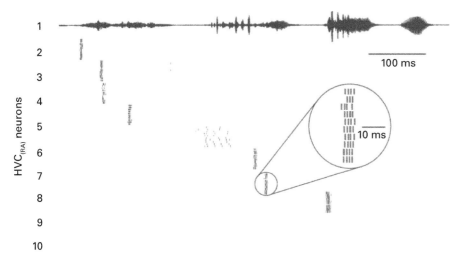

Figure 2.7
Repeating patterns of spikes. *Row 1*, The acoustic recording of a bird singing a well-rehearsed song (time goes from left to right; scale bar gives 100 ms). *Rows 2–10*, Extracellular recordings of spikes from neurons in the bird's high vocal center (HVC) brain region, for 10 repetitions of the song. Each row of tick marks shows the spike generated from one instance of the song. Circular inset shows that neurons fire at nearly the same time on every song, with a precision finer than 10 ms. Adapted from Hahnloser, Kozhevnikov, and Fee (2002).

A Simple Model of Emergent Phenomena in the Brain

To better understand how these emergent phenomena are generated, let us make a very simple network model, inspired by neuroscience, to gain some intuitions. We will vary the way in which the neurons in the model interact with each other to see how that affects the types of emergent phenomena that appear. Although it has few parameters, this model can produce oscillations, synchrony, repeating patterns, and multiple wave types.

The insight that neuronal populations can interact to produce complex emergent patterns was first realized by Wilson and Cowan (1972, 1973). They developed an elegant and powerful set of equations that have proved remarkably useful (Destexhe and Sejnowski 2009; Kilpatrick 2014). Amazingly, the implications of the Wilson-Cowan model are still being worked out in various areas of neuroscience, nearly 50 years later (Inagaki et al. 2019; Roberts et al. 2019). The computer model presented here takes its inspiration from these equations and could be considered a crude digital approximation to them.

Now let us describe the details of this digital model. We will start with a square sheet of 441 neurons (21 rows by 21 columns; figure 2.8A). Looking at this sheet will be like viewing a patch of cortex from above. Like actual neurons, each model neuron can be either on (1) or off (0) and will have a threshold to determine if it should fire or not. In our simulations, typical thresholds will be values from 1 to 5. If a threshold is 3, for example, the neuron will only fire if 3 or more of its inputs are on in the previous time step of the simulation. We will connect each neuron to all its nearest neighbors on the sheet within some distance. For example, if we set that distance to be 1, then each neuron will receive inputs from four neighbors (figure 2.8B). If we set that distance to be 1.5 units, then each neuron will receive inputs from its eight surrounding neighbors (figure 2.8C). Again, like actual neurons, each

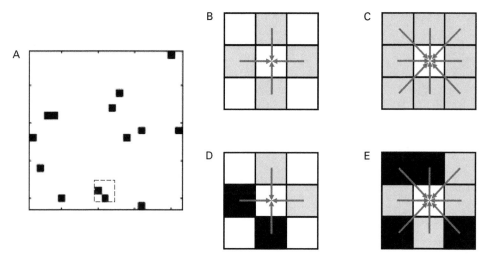

Figure 2.8
A model for studying emergent phenomena. *A*, The model consists of a sheet of neurons arranged in a 21×21 grid. Some active neurons are shown in black from this snapshot of a run of the model. The small dashed square at bottom center highlights the local neighborhood. It encloses nine neurons, two of which are active. *B*, The center neuron will receive inputs (arrows) from only four neighboring neurons (grey) if the maximum connection distance is one unit. *C*, The center neuron will receive inputs from eight neighboring neurons if the maximum connection distance is 1.5 units, allowing the diagonal, corner, neurons to be included. *D*, If the center neuron has a threshold of three, it will not be activated in this situation, as only two inputs are on (black). *E*, If the center neuron has a threshold of three, it will be activated here, as four inputs are on.

model neuron will have a refractory period, indicating how long it will be unable to fire after having just fired. Typical refractory periods will be from 1 to 5. Finally, we will provide some input to the sheet of neurons by setting a probability for each neuron to spontaneously fire. This could be like weak drive from the thalamus coming into the cortex. With a probability of 0.01, for example, about 4 neurons will fire each time step of the simulation ($441 \times 0.01 = 4.41$), even if they are not activated by other neurons. We will explore spontaneous firing probabilities ranging from 0.01 to 0.03. Matlab code for running this model is given through links to the exercises given in the appendix.

If these neurons send enough activity to another neuron to drive it over its threshold, then this neuron will in turn become active. Because the input from one neuron is usually not enough by itself to activate another neuron, activity in the network must be driven by collective interactions among neurons. As we will see, this interesting arrangement allows complex emergent phenomena to exist. In general, activity will propagate if enough neurons are driven over threshold to sustain continued reactivation.

Notice that some parameter values will dampen activity while others will amplify it. A large threshold, a long refractory period, and a short distance over which the neurons are connected, leading to few inputs, will all tend to reduce activity. Conversely, a low threshold, a short refractory period and a large distance over which neurons are connected, leading to many inputs, will tend to increase activity. If activity is largely damped, then we see only the random activations caused by the spontaneous firings that drive the network. On the other hand, if activity is largely amplified, we see explosions of activity, interrupted only by refractory periods. Based on our discussion from chapter 1, we predict the most interesting regime of activity will be somewhere near a balanced state, where activity is neither damped nor amplified.

To start our exploration of the different types of emergent phenomena in this model, let us look at a few representative examples to see if we can recapitulate the phenomena that have been reported to occur in the brain.

The model can generate both round and spiral waves, as shown in figure 2.9. The details of the model settings are given in the figure caption, but we can summarize them quickly. Here the neurons are given many inputs to favor activity, but they are also given a high threshold to discourage activity. This balanced situation leads to interesting activity patterns. Waves in neural media have been extensively studied; for an excellent treatment, see Bressloff (2014).

The model can also generate pulse like waves if we give it slightly different settings (figure 2.10). Here the parameters again are roughly in balance, but with increased random drive, the waves are less likely to remain round or spiral and instead appear as pulses.

To create synchronous activity, but without wavelike structures (figure 2.11), we can scramble the connections of the model so that each neuron is not connected to its nearest neighbors. The settings again promote balance between activity and inactivity. Here, there are random fluctuations until they go over a threshold when they are amplified to produce a synchronous burst.

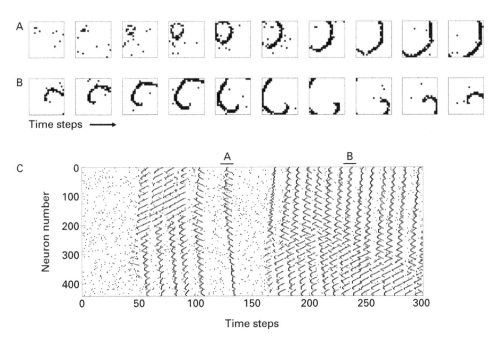

Figure 2.9
Round and spiral waves. *A*, A sequence of 10 frames of activity from the model showing a round wave expanding. *B*, A sequence showing a spiral wave rotating counterclockwise. *C*, A raster plot of activity from the model over time. Neuron number is given on the y-axis, time steps are given on the x-axis. Each dot represents a neuron active at a given time. Random background activity can be seen near time 150, while the round wave occurred at time 123, indicated by the horizontal line at A above the raster. The spiral wave occurred around time 230, indicated by the horizontal line at B. Model settings: a maximum connection distance of 3 units (giving 28 total inputs from neighboring neurons), a neuron threshold of 4, a refractory period of 5, and a probability of spontaneous activity of 0.02.

A

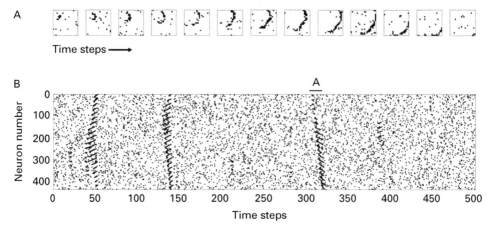

Time steps ⟶

Figure 2.10
A pulse-like wave. *A,* A sequence of 13 frames of activity from the model showing a pulse-like wave, expanding and moving from upper left to lower right. *B,* The raster plot of activity from the model over time. Notice that background activity is more prevalent here, as the probability of spontaneous activity was increased. The pulse wave shown occurred near time 310, as indicated by the horizontal bar under A. Similar pulses, not shown, occurred near times 50 and 140. Model settings: a maximum connection distance of 3 units (giving 28 total inputs from neighboring neurons), a neuron threshold of 5, a refractory period of 3, and a probability of spontaneous activity of 0.03.

A

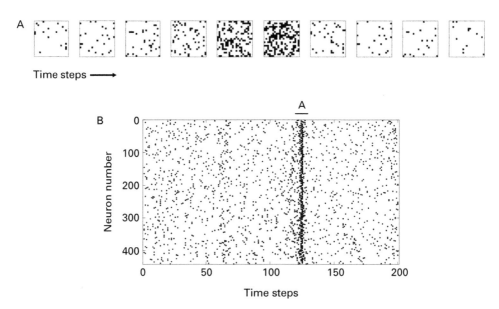

Time steps ⟶

Figure 2.11
Synchronous activity. *A,* A sequence of 10 frames from the model showing a single episode of synchrony. *B,* The raster plot of activity shows the synchronous event occurring near time 125, indicated by the horizontal bar under A. Model settings: a maximum connection distance of 2.5 units (giving 20 total inputs from neighboring neurons), a neuron threshold of 3, a refractory period of 5, and a probability of spontaneous activity of 0.015. Here, the connections were not to nearest neighbors but were randomly assigned to neurons throughout the network.

Just as oscillations with different frequencies occur in the brain, the model can produce oscillating activity with varying frequencies. In these situations, the settings promote excessive activity, but the refractory periods of the neurons determine when this activity will have to turn off. By increasing the refractory period, we can increase the cycle length and reduce the frequency of the oscillations. In this way, the model can produce oscillations at nearly any frequency. Figure 2.12 shows oscillating activity with a period of 5.

The model can also generate repeating patterns of activity that are not wavelike (figure 2.13), just as seen in the brain. In this situation, the model is set as in the previous case where it produced oscillations, but here the connections are not to nearest neighbors, but randomly dispersed throughout the network. This disrupts the wavelike appearance but retains the repeating patterns.

Finally, it is also possible to produce random activity in the model (figure 2.14). For this to happen, the neuron thresholds must be high, to prevent a burst of activity from amplifying until the entire network is saturated. Under high-threshold conditions, there is just the random spontaneous activity occurring in the background.

Complex Emergent Phenomena Occur at a Phase Transition

How can we make sense of this large number of parameters and all the different activity patterns? Is there some way to summarize what is going on here, or must we constantly focus on the details? Fortunately, there is a relatively simple unified picture of this system. If we try all the different combinations of parameters, we notice that there are three general types of activity produced by the model. This is most clearly seen when we plot the

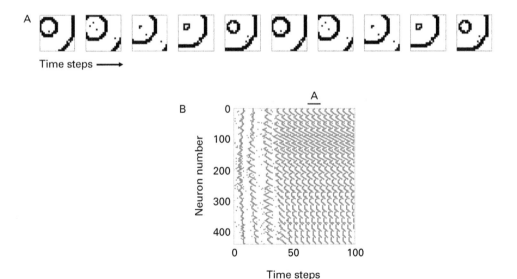

Figure 2.12
Oscillatory activity with period of 5. *A*, A sequence of 10 frames from the model showing an oscillating pattern of waves. Notice that it repeats itself after every five time steps. *B*, The raster plot shows the network falls into this oscillatory pattern and persists, indicated by the horizontal bar under A. Model settings: a maximum connection distance of 3 units, a neuron threshold of 2, a refractory period of 4, and a probability of spontaneous activity of 0.02.

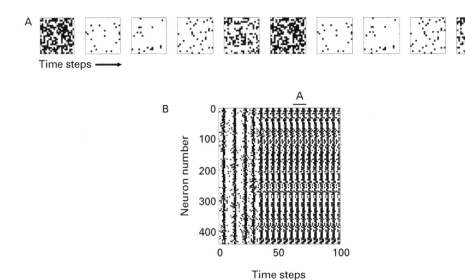

Time steps ⟶

Figure 2.13
A repeating pattern. *A*, A sequence of 10 frames from the model showing a repeating pattern that does not have
a wavelike form. Notice this also repeats itself after every five time steps. *B*, The raster plot shows the network
falls into this repeating pattern and persists, indicated by the horizontal bar under A. Model settings: a maximum connection distance of 3 units, a neuron threshold of 2, a refractory period of 4, and a probability of
spontaneous activity of 0.02. Here, the connections were not to nearest neighbors but were randomly assigned
to neurons throughout the network.

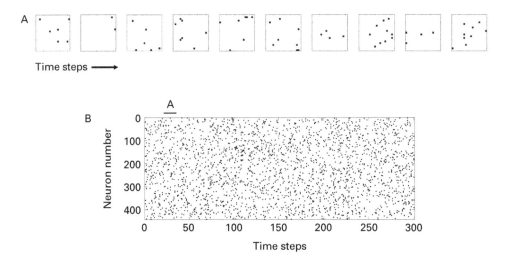

Time steps ⟶

Figure 2.14
Random activity. *A*, A sequence of 10 frames from the model showing random activity. Note the relatively low
density of active neurons, showing that nothing is driven over threshold to elevate activity beyond the random
firing probability. *B*, The raster plot shows disorder. The sequence shown in A above occurred at the time indicated by the horizontal bar. Model settings: a maximum connection distance of 3 units, a neuron threshold of
5, a refractory period of 5, and a probability of spontaneous activity of 0.02.

fraction of neurons that are active across time (figure 2.15). Here we see that the activity ranges from periodic (panel A), to semi-periodic (panel B), to random (panel C). Periodic activity over time is produced by oscillations (figure 2.12) as well as repeating patterns (figure 2.13). Let us call this the ordered type. The semi-periodic activity corresponds to cases where there are round waves (figure 2.9A), spiral waves (figure 2.9B) and pulse-like waves (figure 2.10). It has periods of little to no activity with intermittent bursts of waves that have some periodicity. Let us call this type of activity complex. Finally, random activity has no periodicity to it (figure 2.14). Let us call this the disordered type.

Now if we arrange the different parameter values in a table, with their corresponding types, we notice that most combinations lead to ordered or disordered activity (figure 2.16). Only a few parameter combinations lead to complex activity, and these occur at the border between disorder and order. We can call this the phase transition zone, and it is analogous to the phase transition we saw in chapter 1 when we discussed our first simple neural model. From this perspective, the present model has an ordered phase and a disordered phase, with a phase transition between them. Complexity occurs at the phase transition.

The model has qualitatively reproduced the emergent phenomena seen in the brain identified in our brief summary of the literature. It did this without complex dendrites or diverse ionic currents or even inhibitory neurons. This is not to say that biological details are irrelevant. It is to say, however, that some emergent phenomena do not seem to be dependent on lower-level details, something we have already considered in our discussion of the stadium wave. A very generic model without much biological realism can serve to capture the broad features of emergent phenomena reported in neural tissue.

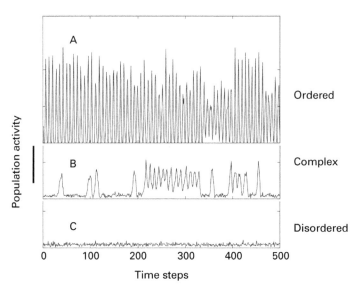

Figure 2.15
Three types of activity. *A*, When thresholds are low, activity explodes and is brought back down by the refractory period. This causes periodic activity characteristic of the ordered type. The fraction of all active neurons, the population activity, is plotted on the y-axis and ranges from 0 to 1. Scale bar height is 0.2, or 20 percent of neurons active. Time steps are on x-axis. *B*, When thresholds are intermediate, activity is complex and is a mixture of periodic and random. *C*, When thresholds are high, random drive does not activate many neurons, and activity is disordered. These three types can be used to broadly classify all the patterns produced by the model.

| | More activity ← → Less activity | | | | | |
	Thresh = 1	Thresh = 2	Thresh = 3	Thresh = 4	Thresh = 5	Thresh = 6
Ref = 1	Ordered	Ordered	Ordered	Ordered	Ordered	Disordered
Ref = 2	Ordered	Ordered	Ordered	Ordered	Disordered	Disordered
Ref = 3	Ordered	Ordered	Ordered	Ordered	Disordered	Disordered
Ref = 4	Ordered	Ordered	Ordered	Complex	Disordered	Disordered
Ref = 5	Ordered	Ordered	Ordered	Complex	Disordered	Disordered
Ref = 6	Ordered	Ordered	Ordered	Complex	Disordered	Disordered
Ref = 7	Ordered	Ordered	Ordered	Complex	Disordered	Disordered

(More activity ↑ / Less activity ↓ on row heads)

Figure 2.16
The phase space of the model. The three types of activity (ordered, complex, disordered) are shown for each combination of threshold (Thresh; column heads) and refractory period (Ref; row heads). Note that the region showing complex activity is relatively small and lies on the border between the ordered and disordered phases. For all entries in this table, the connection distance was set to 3 and the probability of spontaneous activity was 0.02. Increases in the refractory period or in the threshold lead to decreases in activity.

What does seem to matter in this simplified model is how the neurons interact with each other. If there is too much excitation, the activity becomes periodic and ordered. If there is too little excitation, the activity becomes random. Complex activity requires a threshold that is not too high, as that would prevent activation. If the threshold is too low (say 1), then neurons can be fired without collective interactions. In our model, a threshold of 4 promotes the right level of cooperation among neurons to support the emergence of waves, spirals, and pulses.

More complex models with biological realism can generate these emergent phenomena as well. These models range from the extremely detailed simulations of the Blue Brain project, where they attempted to include every known neuron type and current (Markram 2006), to those with moderate realism that simulate multiple, though not all, cell types (Izhikevich and Edelman 2008). A model with intermediate detail can produce phase plots showing different regions of characteristic activity patterns (Brunel 2000). Though the phases in the Brunel model are not exactly the same as those we have in our simple model, they are qualitatively similar. These include a state called "asynchronous irregular" (AI) that is like our disordered phase, and a state called "synchronous regular" (SR) that is like our ordered phase. These phases have a relatively narrow boundary between them called "synchronous irregular" (SI), similar to the region of complex activity we saw between the ordered and disordered phases in our model. Another moderately realistic model produces pulses when excitation and inhibition are balanced, but plane waves and spirals when excitation is increased (Keane and Gong 2015). This again demonstrates that different phases can be reached by tuning the interactions between the neurons, even though the details of these neurons may differ from model to model.

More Complex Emergent Phenomena?

We see that some of the more interesting emergent phenomena like waves, pulses, and spirals occur near the transition between ordered and disordered phases. But why should we care about pulses traveling around in the brain and colliding, or repeating patterns caused by waves? We may be able to account for some basic computations with colliding pulses and perhaps account for memory with repeating patterns, but not much more. How complex can emergent phenomena get and why should we care?

What I have presented up to this point is a necessarily simplified picture of emergent phenomena. The goal has been to derive some intuitions about them. Now that this has hopefully been established, I would like to briefly demonstrate that emergent phenomena can get extremely complex, far more so than what we have been discussing up to now.

To demonstrate this increase in complexity, let us take our simple neural model and change some of the parameters a bit. First, we will eliminate the refractory period, so each neuron can fire at higher rates, and could even be on continuously. Second, we will make the threshold depend on whether the neuron is on or not. If a neuron is off, it will be activated if it receives activity from exactly three neighbors. If a neuron is already on and receives input from two or three neighbors, it will continue to be on. If it is already on and receives input from more than three neighbors, it will turn off. Third, we will not give each neuron a small probability of becoming spontaneously active. Instead, we will just start from some initial configuration with some neurons already on. Aside from these changes, we will keep other things the same. Each neuron will be connected to its eight nearest neighbors, and the neurons will be arranged in a two-dimensional grid. Note that these rules contain some factors that will promote excitation, like a relatively low threshold of two for already active neurons, and no refractory period. They also contain factors that will promote inhibition—the neuron will turn off if it receives input from four or more neighboring neurons, and the neurons will no longer become spontaneously active. Perhaps these rules will set up a roughly even balance between excitation and inhibition. The absence of spontaneous random activity could allow for precise patterns that will not be corrupted by noise.

Well, what does this produce? From most starting conditions, say with about 50 percent of the neurons initially turned on, the activity continues for about 10 to 100 time steps. It appears random at first sight, but upon closer inspection we can find certain patterns of neurons that continue to reappear. Some of these patterns are static, while others move across the screen. An example of a moving pattern is given in figure 2.17, which shows a "glider." This pattern changes every time step for four steps, then repeats the cycle again, except in a slightly different position. The net effect is that the glider moves one neuron down and one to the right after every four time steps.

While the glider is more complex than other emergent phenomena we have seen so far, this is only the beginning. The "glider gun" is a pattern of two oscillating blobs that periodically collide with each other to produce gliders. Every 30 time steps, the blobs give birth to another offspring, and it glides across the screen (figure 2.18). This will continue for as long as the simulation runs.

But this glider gun is Neanderthal when compared to a pattern of 5,120 neurons called "Gemini" that self-replicates and erases its parent pattern after about 33.6 million time

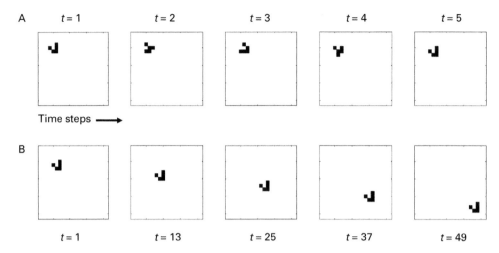

Figure 2.17
A glider. *A*, A configuration of active neurons that repeats itself, but with movement. After four time steps, the pattern is exactly duplicated, but one row lower and one column to the right. This emergent structure is called a "glider." *B*, The glider moves from upper left to lower right. Here, each frame is separated by 12 time steps to show overall trajectory.

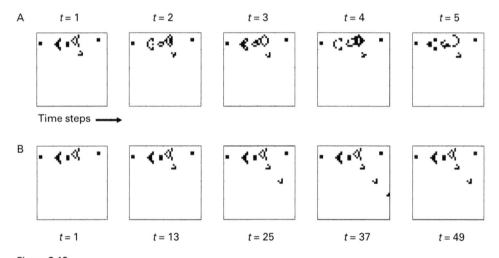

Figure 2.18
A glider gun. *A*, A sequence showing the first five time steps of a configuration known as the "glider gun." Between the two black squares on the upper left and right, there are two collections of neurons that move toward each other and collide in the middle. They change shape as they move, and cycle through these changes precisely every 30 time steps. *B*, When the two groups of neurons collide near the middle, they produce a glider that moves from there to the lower right. Here, each frame is separated by 12 time steps to show overall trajectory.

steps! Interestingly, Gemini performs this replication process by accessing information from a long diagonal tape consisting of gliders (https://conwaylife.com/wiki/Gemini). The similarities to reproducing life forms and DNA are obvious.

In fact, this set of rules is known as the "Game of Life" and was invented by mathematician John Conway in 1970 (Gardner 1970). He was trying to come up with a rule set that promoted complex patterns. Rather than thinking of neurons, he saw each spot on the grid as a potential person. If there was overcrowding with more than three neighbors, then death would occur. If there were three neighbors, but no more, then birth would occur. Again, we see this aim to balance activity and inactivity to produce complexity.

The Game of Life has been intensively studied since its inception, and by now over a thousand different types of complex moving patterns have been discovered (https://www .conwaylife.com/wiki/Main_Page); it remains an active area of research. Using collisions of gliders, researchers have shown the logical functions AND, OR, and NOT can be implemented, proving any type of computation that can be done by a computer can also be accomplished in the Game of Life (Rendell 2011). This means any program, even extremely realistic virtual reality simulations, could be controlled by the Game of Life. The complexity is unlimited, and it would have been nearly impossible to imagine this back in 1970, when only the simple rule set was known. As we previously saw, the types of emergent phenomena that can appear are not easily deduced from just a knowledge of how the small-scale parts will interact.

Researchers have explored slight modifications to the rules of the Game of Life to see how they affect the complexity of the patterns produced. They found the vast majority of rules lead to either (1) chaotic or random behavior, or (2) complete extinction of activity. The minority of rules at the boundary between these two phases produce interesting patterns (Nordfalk and Alstrøm 1996). This echoes what we saw with the phase space of the simple neural model discussed earlier—complexity lies near the phase transition.

Is it possible that structures like Gemini occur in the brain? While the rules in the Game of Life are not too far away from the nearly balanced condition we think prevails in the cortex (Mariño et al. 2005), and while there is evidence the cortex operates near a phase transition point, it seems unlikely that Gemini itself appears in neurons. This is because Game of Life structures are very intolerant of noise. For example, activating a single neuron adjacent to a glider pattern is enough to disrupt it. A single misfiring neuron would therefore clog the information tape used by Gemini, blocking reproduction. We know cortical neurons can give highly variable responses, even to the same stimuli, and synaptic transmission is itself probabilistic; there are plenty of sources for noise (Faisal, Selen, and Wolpert 2008). This makes it seem ridiculous to think of patterns like Gemini appearing in the brain.

Why then talk about these things? There are at least three reasons. First, even though the brain has many stochastic components, its output can often be highly reliable. This is because large numbers of neurons can compensate for the variability of individual neurons. For example, if 100 neurons spike to encode the direction in which an arm is to move, a weighted population average can represent the direction much more accurately than most individual neurons alone (Georgopoulos, Schwartz, and Kettner 1986). Population codes could therefore correct errors and allow very intricate emergent phenomena to exist in the brain. Second, the rules in the Game of Life are relatively simple and it is quite likely that the set of rules governing activity in the cortex are richer. These could allow for equally interesting structures, different from Gemini, to emerge. It is quite possible that language, complex reasoning, and consciousness are all products of these rules, whatever they are. Third, from what we have seen, it seems likely that these rules will lie near a phase transition within their rule space, just like Gemini does. An understanding of where the complex phenomena occur, even if they are exotic and unlikely to appear in the brain, may still generalize and tell us something about where to look for them in other systems.

How to Study Emergent Phenomena

In closing this chapter, we can now make a few remarks on how emergent phenomena should be studied, and where we should exercise caution. This is relatively new ground, given the historical dominance of patch clamp techniques and fMRI, which have focused on the micro and the macro scales, respectively. Most experiments are not designed with the meso scale in mind, so emergent phenomena have often been overlooked, and rarely quantified (Daniels et al. 2016). How can we study them?

1. Observe as many interacting parts as possible. Because emergent phenomena are produced by the interactions of many constituent parts, it is important to simultaneously observe as many constituent parts (e.g., neurons, brain regions) as possible. Observing only four people in a stadium would give us no clue about the wave; at least 100 would be required to even begin to suspect an emergent event. It is likely to be the same with neurons.

2. Use appropriate spatiotemporal sampling. To understand the interactions of many parts, it is essential to observe them at the natural timescale of their interactions. For example, in the cortex this would mean simultaneous recordings from hundreds of synaptically connected neurons at a temporal resolution of around 1 ms. The distance of nearest-neighbor synaptic connections in the cortex is within about 100–200 µm (Holmgren et al. 2003; Perin, Berger, and Markram 2011), and synaptic delays between nearby pyramidal neurons are usually less than 2 ms (Mason, Nicoll, and Stratford 1991). While not all recording methods can attain these requirements, many are almost there. Methods that subsample in space may fail to record from synaptically connected neurons, and methods that subsample in time may artificially lump together multiple independent events that are only approximately coincident.

3. Identify connectivity. Because emergent phenomena depend on the interactions of constituent parts, it is important to identify the pattern of connectivity among the parts. For the cortex, this would mean a map of functional connectivity among the recorded neurons. Which neurons are nearest neighbors from the perspective of propagating activity? Which neurons are interacting?

4. Manipulate interactions. Because these phenomena emerge as a result of how their constituent parts interact, it is important to manipulate the interactions, if possible. In neural systems this can come from changing the balance of excitation and inhibition or by altering levels of neuromodulators. With a behaving animal, different states like attention and sleep can alter interactions between neurons. In simulations, targeted manipulations can of course be more extensive and include changes in connectivity, the proportion of excitatory and inhibitory neurons, the refractory period, the thresholds, and other factors that would be difficult or impossible to manipulate in living tissue.

5. Make a phase diagram. The results of such manipulations will typically produce changes in characteristic behaviors (e.g., oscillations versus no oscillations, ordered activity versus no activity, amplified versus damped activity). These characteristic behaviors should be identified and used to construct a phase diagram. Special attention should be paid to boundaries between phases, as the transition region often harbors the most subjectively complex and interesting forms of activity.

6. Observe dynamics in the phase diagram. Given the phase diagram, it is important to track how the system spends its time there. Does it usually reside near the phase transition, or is it consistently deep in one of the phases? It is useful to identify things that move the system around in the phase diagram. For example, does it move from one phase to another as the organism transitions from sleep to wakefulness? When the organism is in focused attention, does it move more toward an ordered phase? Does the system show homeostasis? When it is perturbed from a location in phase space, does it return there after some time? If so, is such a return merely the result of drift, or are there mechanisms actively moving it there? Is the system operating to consistently return to some set point?

7. Look for causes at multiple scales. Conclusions drawn from manipulations to constituent parts should be interpreted carefully and not overstated. For example, an experiment may show that REM sleep is eliminated after a gene is knocked out. This tells us that the gene is somehow involved in REM sleep, or supports it, but it does not tell us that this gene, by itself, causes REM sleep and is the sole explanation for it. This is because REM sleep is an emergent phenomenon, produced by the way in which many neurons interact. A full understanding of REM sleep would be expected to involve many different levels. Any explanation that stops at the smallest scale is likely to be incomplete.

Next, we will revisit the conceptual model introduced in chapter 1, but with quantitative measures. Let us put numbers on our intuitions. This will position us to later understand the data in the current literature and how it is being interpreted.

Chapter Summary

Methodological reductionism has emphasized the smallest scales as most fundamental. While this has been very successful, it has failed to illuminate our understanding of emergent phenomena. Neural waves and oscillations arise out of the interactions of vast numbers of neurons. They have properties different from the neurons that make them up, and they occur at larger scales. The interactions between neurons can control the types of characteristic phenomena that emerge. Complex emergent phenomena can often be found near a phase transition point. The complexity of emergent phenomena there is virtually unlimited, allowing universal computations. To experimentally access these phenomena, it will be helpful to develop more holistic approaches. The vast world of complexity can be accessed through the narrow zone of the phase transition, near the critical point.

Exercises for this chapter can be found through a link given in the appendix.

II

The Critical Point and Its Consequences

3

The Critical Point

Bottomless wonders spring from simple rules that are repeated without end.
—Benoit Mandelbrot

In the basic neuron-like model of the last chapter, the most interesting emergent phenomena appeared near the phase transition point. We saw that although the rules were simple to state, they could produce great complexity when the simulation was tuned to this phase transition—the model replicated round and spiral waves, traveling pulses, oscillations, and repeating patterns. These emergent phenomena are seen in the brain and have been reported by many research groups.

Now I want to turn to the phase transition point itself and see if another simple model can accurately predict what the data should look like there. We will pick up the branching model we first discussed in chapter 1 and describe it in more detail here. Instead of looking for waves and emergent phenomena, I want to see what the signatures of the critical point are in the branching model. We know that it should produce power laws, as we mentioned earlier, but we want to see more than that now. Can we reveal several additional signatures of the phase transition by examining this model? If we can, then these will serve as predictions we can test in the physiological data.

Why is the critical point so important, and why devote an entire chapter to it? One of the main contentions of this book is that networks of cortical neurons operate near the critical point. During development, they wire themselves up so that they gradually approach it as they mature. When they are perturbed away from it, networks find a way to get back there through adaptation or homeostasis. And when they are near the critical point, many of their information processing functions are optimized. Before we go further, then, it will be crucial to define the critical point more quantitatively and to examine the data to see if they show the expected signatures of being near it.

I also want to explain more about why we will be using the branching model for this exploration of the critical point. While there is no shortage of models describing the critical point in neural networks, it is easier to explain things if we focus largely on one model. The branching model is fairly simple, as universality would require, it has been widely used (Zapperi, Lauritsen, and Stanley 1995; De Carvalho and Prado 2000; Haldeman and

Beggs 2005; Kinouchi and Copelli 2006; Pajevic and Plenz 2009; Kello 2013), and as we will see, it fits the data quite well.

But there is one other reason for choosing this path also. The branching model is a non-equilibrium model, and this is essential if we are to capture accurately the unbalanced condition of the cortex. By unbalanced, I mean that each neuron is constantly receiving tens to hundreds of synaptic impulses per second. Further, the rates of these impulses are often changing with stimulus conditions and cognitive state. These facts call for nonequilibrium models that are specifically designed to address this external drive. In contrast, equilibrium models assume that the system to be modeled is closed, not receiving external inputs, and settling into a quiescent state. While equilibrium models might serve well as approximations for some limiting case of low external drive, if we are to realistically deal with the situation that prevails upon brains, nonequilibrium models will be more appropriate.

In the rest of this chapter, then, we will first use the branching model to explore multiple signatures of operating near the critical point. Second, we will treat these as predictions to be tested, so we will review the relevant experimental data to see how well the branching model can do. Finally, we will briefly consider some objections that have been raised about these signatures of criticality and how they should be interpreted.

The Branching Model: A Branching Ratio Near 1

In the branching model, neurons are represented by nodes that can be either on (1) or off (0). Each node can be connected to other nodes via directed connections, and each connection has a transmission probability associated with it. A neuron will fire if one or more of its incoming connections becomes active. For example, consider two active neurons, neurons A and B, each sending one connection with a probability of 0.25 to a third neuron, neuron C. Neuron C will become active if either one of these connections transmits. The computer program would simulate this by drawing a random number between 0 and 1 for each connection. If the random number was below 0.25, then the connection would transmit. If a neuron is not active, then it will not transmit, and no random number is drawn. Notice that neuron C does not have a threshold, and it is possible that just one connection will transmit and cause it to fire. In this way, the branching model differs from actual neurons—there is no threshold requiring, say, two presynaptic inputs to be simultaneously active to fire the neuron.

In all the simulations that we will discuss here, each neuron is given a fixed number of outgoing connections that it can make to other neurons. In the examples that will follow, these connections are assigned randomly to other neurons, creating what is known as a random graph. It is possible to connect them in structured ways, though, to simulate small-world or scale-free structures. Much later, we will discuss how particular, nonrandom network structures can influence the dynamics. But for now, the random pattern of connectivity can serve to illustrate the most important results.

We can tune the branching model by setting the sum of the outgoing transmission probabilities from all the nodes. In figure 3.1, for example, the node has four connections whose probabilities sum to 1.0. The sum of the outgoing probabilities will determine the branching ratio *by construction*. This will allow us to make the model critical when the sum equals exactly 1, or subcritical when it is less than 1, or supercritical when it is more than 1.

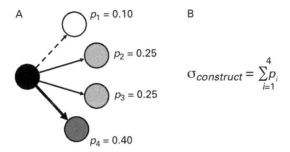

A

$p_1 = 0.10$

B

$p_2 = 0.25$

$$\sigma_{construct} = \sum_{i=1}^{4} p_i$$

$p_3 = 0.25$

$p_4 = 0.40$

Figure 3.1
Constructing the branching ratio σ for a model. *A*, A neuron with four outgoing connections of different transmission probabilities. These will control how activity spreads in a network. For example, when the neuron on the
left is on (black), it has a 40 percent chance of activating the lower neuron on the right in the next time step
($p_4 = 0.40$). *B*, The branching ratio, σ, is set by construction to be the sum of the outgoing transmission probabilities. Here, they add to 1.00, poising this neuron at the critical point. By constructing a network where all the
neurons have a given branching ratio, we can control whether the activity is damped ($\sigma < 1$), preserved ($\sigma = 1$) or
amplified ($\sigma > 1$).

Because this is a nonequilibrium model, there must be some external drive to activate
it. We typically use a very small probability, say $P_{spont} = 10^{-5}$, to model the spontaneous
firing of each neuron. That means that every time step, a random number is drawn for
each neuron. If that number is smaller than 10^{-5}, then that neuron will become active,
regardless of whether it receives inputs from other neurons or not. In this manner, the
network is constantly receiving a weak drive. When the model is run for hundreds of
thousands of time steps, even this small probability causes enough avalanches to form to
generate a relatively broad distribution of avalanche sizes.

Once so many avalanches are produced, it is then possible to measure the branching
ratio *empirically*. This can be done very crudely by simply observing the average number
of descendant neurons that are activated whenever there is one active ancestor neuron
(figure 3.2). Such rough estimates were used in early experimental work on neuronal avalanches (Beggs and Plenz 2003), but since then a more sophisticated and accurate method
has been developed (Wilting and Priesemann 2018). Briefly, this new method looks at the
activity level over many time steps and notes whether it is gradually declining, holding
steady, or gradually increasing. It then finds the best fit curve to the data and from there
calculates the branching ratio that could have produced that curve.

Why do we need to empirically observe the branching ratio from the output when we
can set it by construction in the model? Shouldn't the observed branching ratio always
agree with the one that was built into the model? While these usually agree pretty well,
they sometimes don't, and it can be illuminating for us to know when that occurs. For
example, let's say we constructed a branching model to be critical, so that the sum of outgoing transmission probabilities from every neuron was 1. We run the model but find that
the observed branching ratio is only 0.95. What happened? It turns out that high spontaneous firing probabilities can cause the observed branching ratio to decline. In this case,
high spontaneous activity causes some neurons to fire by themselves, without creating
any descendants. The increased number of avalanches of size and length one reduces the
branching ratio. As the spontaneous firing probability approaches zero, the discrepancy
between the constructed and observed branching ratio becomes less. Given relatively low

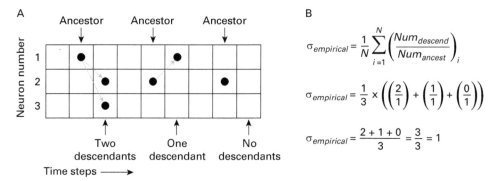

Figure 3.2
Measuring the branching ratio σ from data. *A*, Raster plot shows activity from three neurons over time. Each avalanche is composed of one ancestor neuron and two, one, or zero descendants. *B*, The empirical branching ratio σ is calculated as the average number of descendants produced by one ancestor. The method illustrated here is to provide intuition; more sophisticated methods have recently been developed that can more accurately estimate σ even under conditions of subsampling (Wilting and Priesemann 2018).

noise, then, an empirically measured branching ratio close to 1 can be taken as an indicator that a network is operating near the critical point. This is a signature we will look for in experimental data later.

The Branching Model: A Phase Transition with Control and Order Parameters

Now let's move on to examine another signature of operating near the critical point. If this is to be a model of a network that can operate near a critical phase transition, then we should be able to demonstrate distinct phases, as well as a boundary between them. How can this be done? A naïve but reasonable guess would be that driving the model harder would cause it to become supercritical, while driving it weakly or not at all would cause the model to be subcritical. This is not the case, however. The constructed branching ratio, and not the amount of drive, controls the phases. For this reason, the branching ratio is called the *control parameter*. In figure 3.3A, we plot the activity against the branching ratio for a simple branching network model. Here, activity is measured as the fraction of neurons that are on (which could range from 0 to 1), which we will call the density of active sites, ρ. The density of sites is the *order parameter*. Notice that ρ starts to climb near a branching ratio of 1. To the left of this is a phase where there is no activity and to the right of it is a phase where there is abundant activity. Right at a branching ratio of 1, activity is barely present.

We are now able to describe an important distinction in phase transitions. What we have just shown is called a *continuous phase transition* because the slope of the order parameter, ρ, changes smoothly as we move from one phase to another. In the branching model we have a continuous phase transition; most critical models applied to the brain are of this type (Haldeman and Beggs 2005; Levina, Herrmann, and Geisel 2007; Stepp, Plenz, and Srinivasa 2015; Del Papa, Priesemann, and Triesch 2017). In nature, there is a continuous phase transition between steam and liquid when water is at the right temperature and pressure is brought to the critical point. A similar continuous transition occurs when a hot piece of iron gradually cools in the presence of a magnetic field, becoming magnetized.

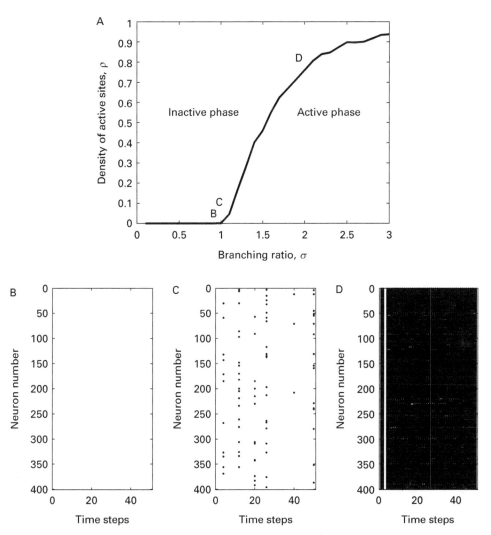

Figure 3.3
A phase diagram for neural network activity. *A*, The density of active sites, ρ, is plotted against the constructed branching ratio, σ, for simple feed forward networks. Notice that the fraction of active neurons, given by the order parameter ρ, jumps up when the control parameter σ is at 1.0. This indicates the critical point, separating the two phases. *B*, *C*, *D*, Sample raster plots of activity when σ=0.9, 1.0, 2.5, respectively. To the left of the critical point, there is a phase of no activity (subcritical), to the right of it there is a phase with activity (supercritical). Network model had 400 neurons in each layer and 60 layers.

In contrast, at a *discontinuous phase transition* we would see a sudden jump in the order parameter at the critical point. While there are not many models of the brain that propose this type of transition, there are a few (Millman et al. 2010; Martinello et al. 2017; Scarpetta et al. 2018), and it is an open question which type of transition, if any, best describes the neuronal data. In nature, a discontinuous phase transition is seen when water melts from ice to liquid. For historical reasons, discontinuous phase transitions are often also called first-order phase transitions and continuous phase transitions are called second-order phase transitions.[1]

To review this section, the branching model predicts several features relevant to the critical point that we should expect to see in physiological data. Networks of cortical neurons should exhibit distinct phases, like an inactive phase and an active phase. Additionally, these networks should be tunable so that they can be moved from one phase to another. As we learned earlier, the critical point lies between these phases.

The Branching Model: An Exponent Relation between Multiple Power Laws

In chapter 1, we saw that when the branching model was tuned to be near the critical point, it would produce a distribution of avalanche sizes that followed a power law (figures 1.6 and 1.8). In addition to this, though, there should be multiple power laws near the critical point, and there should be an exponent relation between the slopes of these various power laws. In the case of the branching model, we have an exponent for the avalanche size distribution, τ, and for the avalanche duration distribution, α. There is a third power law that relates the avalanche size to its duration (figure 3.4C). We will explain this more soon, but for now we note that it has another exponent, γ. If the system is operating near the critical point, these three exponents can be related to each other through a simple algebraic equation:

$$\frac{\alpha - 1}{\tau - 1} = \gamma.$$

The derivation of this equation, as well as the justification for why it can be satisfied only near the critical point, is given in the appendix (Sethna, Dahmen, and Myers 2001; Scarpetta et al. 2018). Here, we will note that data from critical branching model simulations, with a constructed branching ratio of 1.0, satisfy this equation within relatively small errors (figure 3.4). If the model is constructed with a branching ratio of 0.75 or 1.25, the size and duration distributions will be curved and no longer power laws, and even slopes that come from attempted fits to them will not satisfy the exponent relation. Because these observations are consistent with the theoretical justification, we can consider this another valid indicator of proximity to the critical point. As we will soon see, this exponent relation is also a prediction that can be tested in biological experiments.

The Branching Model: Fractal Copies of Avalanches

Yet another signature to distinguish a critical system is related to the property of being scale-free. To appreciate this, let us first explain what "scale-free" means with a little more mathematical detail. A power law is a description of a scaling relationship. For avalanches, this says that a given ratio of avalanche sizes will have a corresponding ratio of avalanche probabilities. To see this, let us consider two avalanches of different sizes, S_1 and S_2; the ratio of their sizes will be some number: $a = S_2/S_1$. What will be the ratio of their probabilities? We know that $P(S_1) = kS_1^{-\tau}$ from the avalanche size distribution equation, where k is a constant, so it must also be true that $P(S_2) = P(aS_1) = k(aS_1)^{-\tau}$, since $S_2 = aS_1$. Then the ratio of the avalanche probabilities would be $P(S_2)/P(S_1) = k(aS_1)^{-\tau}/kS_1^{-\tau} = ka^{-\tau}S_1^{-\tau}/kS_1^{-\tau} = a^{-\tau}$. This number, $a^{-\tau}$, does not change with avalanche size, and only depends on the ratio of the sizes of the two avalanches being compared. Thus, the ratio of probabilities of occurrence

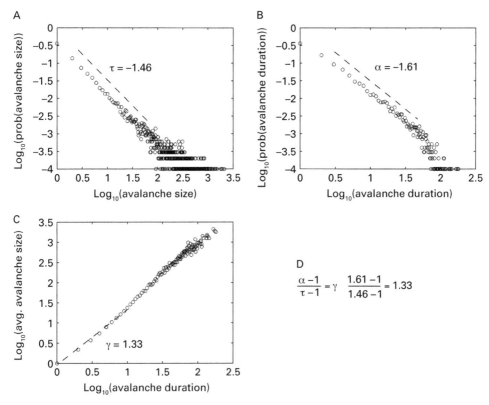

Figure 3.4
Satisfying the exponent relation. *A*, Avalanche size distribution from a branching model shows a nearly power-law distribution. The exponent of 1.46 was determined by a software package for fitting power laws. Model had 128 neurons and a constructed branching ratio of 1.0. *B*, Avalanche duration distribution was fitted and had an exponent of 1.61. *C*, Average avalanche size for a given duration is related to duration via another power law. Here, a line with exponent 1.33 was drawn to the beginning portion of the plot for inspection. *D*, Despite the small model size, these exponents approximately satisfy the relation, indicating closeness to the critical point. Software for fitting power laws is from Marshall et al. (2016).

between an avalanche of size 10 and one of size 1 is the same as the ratio of probabilities of occurrence for an avalanche of size 10,000 and one of size 1,000. The distribution preserves this ratio, regardless of avalanche size. In this sense, it is independent of scale and is therefore called "scale free." For any point along the power law, if you were to go down or up in size by the same interval, numbers would always be scaled down or up by this same factor.

When a system is near the critical point, as we have seen, it will produce many relationships that have this scale-free property. In contrast, a system without a phase transition may have one power law but will not have several. When near the critical point, this scale-free property is quite broad and should extend beyond power laws, which are merely relationships between two variables (e.g., *S* and *T*). If we can find some feature of the avalanches in addition to their sizes and durations, that feature should also display scale invariance. It turns out that each avalanche can be characterized by its temporal profile, or shape. By this, we simply mean the number of neurons that were active at each time step, as shown in figure 3.5. This shape could conceivably be relatively flat for short avalanches (the same number of neurons active in each time step) and very peaked for large avalanches (more neurons

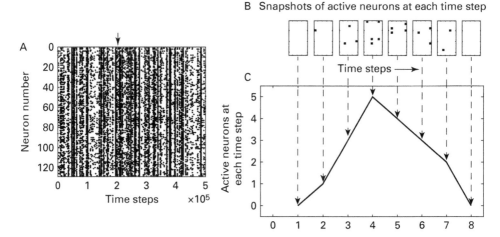

Figure 3.5
Avalanche shape. *A*, Raster plot of activity from a branching model with 128 neurons. *B*, One avalanche, taken near the time marked by the arrow in A, is shown as a series of eight frames. Each frame shows the neurons arranged in a 16×8 sheet, where active neurons are in black. *C*, The shape of the avalanche is the profile produced when the number of active neurons in each frame are plotted across time. On average, avalanches start with a small number of spikes, then have an increase in spikes toward the middle, followed by a decline in spikes toward the end.

active in the middle of the avalanche than at the ends), while still holding to the power law relating size and duration: $S = T^\gamma$. But if the average avalanche shape is preserved across sizes, such that each shape is just a scaled copy of all the other shapes, then we would have an example of a high-dimensional feature that exhibits the scale-free property. This would be more convincing evidence that the system is operating near the critical point.

When avalanche shapes from the branching model are averaged for a given duration, they display the shape of an inverted parabola, as shown in figure 3.6. Interestingly, this shape is preserved for short and long avalanches, suggesting that there is a single shape that is scaled up and down, depending on avalanche length. The factor by which this scaling occurs is connected to the exponent γ which relates avalanche duration to avalanche size. If we had a model of infinite size, the average avalanche shape for every avalanche would follow this inverted parabola. For this reason, the curve is called a "universal scaling function." Note that a universal scaling function does not necessarily have to be a symmetric, inverted parabola—it could be a skewed parabola or a semicircle. The main idea here is that whatever the shape is, it appears the same across multiple scales. In the appendix, there is an explanation of how the exponent γ is involved in the scaling of avalanche shapes to this universal scaling function.

Of course, in finite models we will only be able to see avalanches adhere to this shape within a limited range. In our simulation here, for avalanches of length three, there will be three points, so a tent-like shape will only approximate the parabola. Avalanches of length 15 and greater occur so rarely that their average will be noisy and not form a smooth parabola. This could be overcome by running a much larger model for a much longer time. Even so, the limited model we have here is enough to demonstrate that a universal scaling function is a high-dimensional example of the scale-free property. This is the sixth signature that the system is operating near the critical point and is an additional prediction that we will look for in the physiological data.

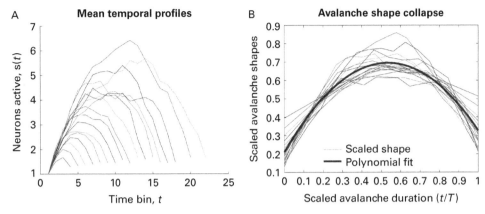

Figure 3.6
Avalanche shapes are rescaled copies. *A*, Average avalanche shapes for durations from 4 to 22 plotted on the same scale. Similar shapes are apparent, despite different sizes. *B*, When these avalanche profiles are rescaled, they approximately collapse on top of each other, showing similarity in shapes. To rescale widths, each shape was divided by its duration, *T*, so that all avalanches had a length of one. To rescale heights, the number of active neurons at each time step was divided by $T^{\gamma-1}$. Branching model data taken from the same simulation that produced figure 3.4. Software package for performing this collapse is from Marshall et al. (2016) and a link to it can be found in the appendix.

Signatures of Being near the Critical Point

We can now summarize the various hallmarks of being near the critical point. Although these were demonstrated with the branching model, they are quite general and apply to other models also. Many integrate and fire neural network models would show these signatures when operating near the critical point:

1. The empirically measured branching ratio is near the critical value of 1.

2. A control parameter tunes the system, transitioning it from one phase to another.

3. The order parameter indicates distinct phases on either side of the transition.

4. Near the critical point, several power laws appear.

5. There is an exponent relation between these multiple power laws.

6. There is a universal scaling function (e.g., scaled copies of avalanches).

As we will see later in chapter 4, information processing functions that show a sharp peak at the critical point can be considered another signature of criticality. But for now, let's see how these six signatures apply to experimental data.

Signatures of the Critical Point from the Data

For about the past 20 years, a wide variety of species and methods have been used to examine claims of criticality in the brain. With so many different approaches, it will be useful to focus our assessment by starting with just one type of data. Where should we begin? The smallest level at which claims of a critical point in the cortex could be examined would be in a network containing a hundred or more spiking neurons. Neural network models predict that signatures of a phase transition will first arise in networks of this size. Such networks are also the fundamental building blocks that make up larger, more coarse-grained signals like local field potentials (LFPs), electroencephalograms (EEGs), magnetoencephalograms (MEGs) and functional magnetic resonance imaging (fMRI). If we can first grasp how the critical point occurs in networks of spiking neurons, this will give us the foundation to see how it appears at these larger scales. But before we jump in to search for signatures of the critical point in spiking networks, let us describe how these data are collected, as well as the advantages and disadvantages of this approach.

In Vitro Experiments

All the experiments we will look at in this chapter were performed in vitro, meaning "within the glass." In other words, these recordings were made in small glass chambers holding tissue slices from brains or cultures of neurons. While this technique is widely used, many neuroscientists prefer experiments to be done in vivo, "within the living organism." This preference seems natural because in vivo work occurs in the intact brain, often while the animal is freely behaving or at least interacting with its environment by pressing a lever or watching a stimulus appear on a screen. If we ultimately seek to understand how the brain works in governing behavior, then why would anyone ever use small samples of tissue with their many severed connections, sitting in a dish, removed from neuromodulators and without any realistic stimuli? It seems like a poor scientific strategy. Yet many labs, including our own, continue to have a strong in vitro component of research. Are there any good reasons for this? I think there are.

First, the history of science shows that simplified model systems can provide valuable insights when problems are highly complex. Gregor Mendel used pea plants, not humans, to reveal the laws of inheritance. Niels Bohr helped launch quantum mechanics by analyzing hydrogen with its single electron. Xenon, with its 54 electrons, would have been a disaster. Eric Kandel made seminal contributions to learning and memory by studying mere reflexes in the lowly sea slug Aplysia. In contrast, the cellular basis of learning language in humans is still largely unknown. Clearly, science proceeds by having a diverse research portfolio, and it is wise to perform some experiments under realistic in vivo conditions. But given that the brain is so incredibly complex, why not have some research dedicated to in vitro model systems as well?

Second, in vitro preparations allow excellent access to the tissue for control and recording for long durations (figures 3.7 and 3.8). This access permits rapid exchange of pharmacological solutions, stimulation of individual neurons, and very dense electrode arrays that would be difficult to achieve in vivo. In chapter 2, we mentioned several requirements for studying emergent phenomena in living neural networks. Among the requirements was the ability to record from large numbers of neurons simultaneously, the ability to record from neurons that are likely to share synaptic connections, and the importance of high temporal sampling rates to ensure the directionality of propagation can be resolved. Most connections from pyramidal neurons in cortex are made within a radius of about 100–200 μm (Holmgren et al. 2003; Perin, Berger, and Markram 2011). This short distance also has consequences for synaptic delays among closely spaced neurons—their mean value is 1.2 ms (Mason, Nicoll, and Stratford 1991). To observe emergent phenomena then, it is essential to have many densely packed recording sites, sampling at rates 10,000 Hz or higher.

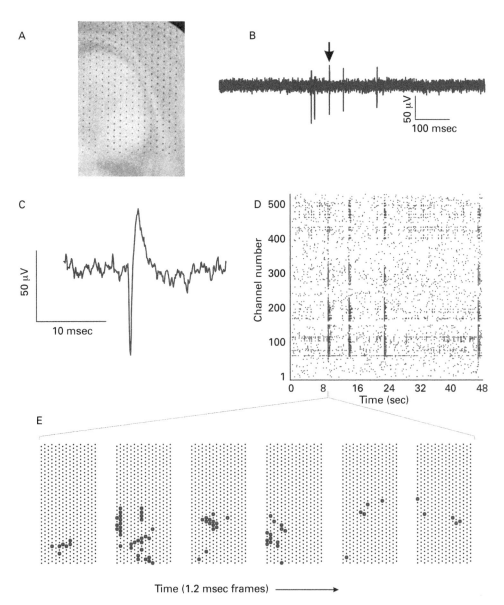

Figure 3.7
Spontaneous spike data collected from a 512-electrode array. *A*, Photo of the slice culture on the 512-electrode array. Electrode tips are small black dots. *B*, Trace from one electrode showing spikes. The arrow points to the spike shown in C. Note that the spike is clearly distinguishable from the background noise. *C*, Spike at higher resolution, showing biphasic waveform and narrow width. *D*, Raster plot of activity from all 512 electrodes over 48 s. Each dot represents a spike. There is nearly continuous background activity with some synchronous periods near 10, 14, 22, and 46 seconds. *E*, A sequence of frames showing spiking activity on the array. Each frame captures activity in a 1.2-ms bin, and larger dots indicate spikes. In this recording, activity appeared on hundreds of electrodes and continued for over 7 hrs.

Electrode arrays implanted in vivo may have individual shanks with closely spaced electrodes along the shanks (30 μm interelectrode distances), but the shanks themselves may be separated by 200 μm or more; a relatively large distance. While in vivo calcium imaging allows large numbers of closely spaced neurons to be recorded, the temporal sampling rate is often insufficient to identify which of two nearby neurons fired first. In vitro experiments with densely packed electrode arrays overcome these limitations. They also permit extremely long duration, stable recordings of 10 hours or more (Beggs and Plenz 2004; Gal et al. 2010). The statistics on neuronal avalanches under such conditions are superb, allowing millions of events to be recorded from hundreds of neurons without perturbations. This is very hard to match in vivo, where incoming stimuli, animal movements, circadian rhythms, and changes in body temperature can alter activity. Taken together, these limitations make it attractive to use in vitro microelectrode arrays for studies of emergent phenomena at the local network level.

Third, it turns out that in vitro preparations clearly exhibit emergent phenomena. Waves (Sanchez-Vives and McCormick 2000), oscillations (Fisahn et al. 1998), synchrony (Cappaert, Lopes da Silva, and Wadman 2009), repeating activity patterns (Rolston, Wagenaar, and Potter 2007; Beggs and Plenz 2004) and neuronal avalanches are all present in these reduced preparations. Interestingly, neuronal avalanches were first discovered in vitro (Beggs and Plenz 2003), prompting the search for them in vivo, where they were later found

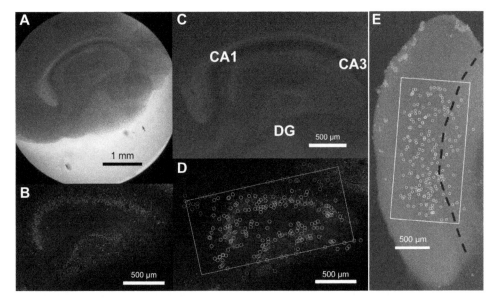

Figure 3.8
A 512-microelectrode array for high resolution recordings of neural networks. Photographs of cortico-hippocampal slice cultures. *A*, Image of an example organotypic culture after 1 day in vitro. The hippocampal structure is visible without staining. *B*, Stained neurons visible in the culture after data collection and tissue fixation. There are some missing neurons in CA3 as consistent with a previous report (Zimmer and Gähwiler 1984), but the overall layer structure is well conserved. *C*, Overlaid photograph of A and B. Positions and dimensions of the hippocampal structures are well conserved during the incubation period. *D*, Overlaid photograph of B, the outline of the array (rectangle), and the estimated locations of the recorded neurons. Light circles are hippocampal neurons identified through triangulation of waveforms from multiple electrodes. Note that locations of the recorded neurons match with the granule cell layer and the cell body layer. *E*, Rectangular outline of array superimposed on organotypic cortex culture. Circles show recorded neuron locations; dashed line indicates border between layers 5 and 6. Figure adapted from Ito et al. (2014).

(Petermann et al. 2009). In addition, the lognormal distribution of firing rates observed in vivo (Hromádka, DeWeese, and Zador 2008) is also seen in vitro (Shimono and Beggs 2015). Apparently, the self-organizing mechanisms present in vitro are robust and general enough to produce all these complex phenomena, even without sensory stimulation or neuromodulators like those that exist in the intact brain.

Now that we have discussed some of the issues surrounding in vitro preparations, let us move on to see what types of data experiments with them produced. These data will help us to assess whether there is good evidence for claiming that the cortex operates near a critical point.

Data: A Branching Ratio near 1

If a network is near the critical point, activity should be neither damped or amplified, but preserved. As we saw, the branching ratio gives the average number of neurons activated by a single active neuron. When this ratio, σ, is equal to 1, the network is exactly at the critical point. Therefore, another piece of evidence to weigh in our assessment of the criticality hypothesis would be empirically accurate measurements of the branching ratio σ. Are they near 1?

Viola Priesemann and Jens Wilting have developed a very precise method for estimating σ even under conditions of subsampling, where activity from only a few neurons is available (Spitzner et al. 2020; Wilting and Priesemann 2018). They have rigorously validated this method on models, showing it to be highly reliable. When they applied it to data sets of spiking activity recorded in vivo from monkeys ($n = 8$), cats ($n = 1$) and rats ($n = 5$), they found the average value to be $\sigma = 0.9875 \pm 0.0105$ (Wilting and Priesemann 2019b).

When we applied their methods to our in vitro data, we obtained similar results. For example, the cortical slice culture shown in figure 3.11 had $\sigma = 0.9971$, which is representative of our data. We have also measured the branching ratio in hundreds of recordings from dissociated cultures (Timme et al. 2016b). In these preparations, neurons are first enzymatically dissolved from tissue and suspended in a solution. This solution is then poured out onto an electrode array, where the neurons attach. Over a period of days, they extend axons and dendrites, forming synapses through self-organization; they then start to fire spikes. The resulting network of dissociated neurons therefore does not have six layers, like what is seen in organotypic slice cultures of cortex or in the intact cortex. Nevertheless, dissociated cultures still produce neuronal avalanches that follow power law distributions (Pasquale et al. 2008), demonstrate avalanche shape collapse (Friedman et al. 2012; see supplemental information), and have a measurable branching ratio. The results of measuring the branching ratio in these cultures can be seen in figure 3.9.

Does the branching ratio ever deviate much from the critical value of 1? As you can see from the histogram, it is sometimes naturally slightly above 1, although this is quite rare. It is possible to make the branching ratio drop to values near 0.5 when pharmacological agents like DNQX (6,7-dinitroquinoxaline-2,3-dione, which blocks excitatory glutamatergic transmission) or muscimol (an inhibitory agonist) are applied. Randomly shuffling the data also causes the branching ratio to drop. These manipulations tell us that a branching ratio very close to 1 requires the proper balance between excitation and inhibition, and that $\sigma \approx 1$ is not what would be expected by chance. These facts also suggest that the brain has mechanisms for maintaining the branching ratio close to 1. This is a subject to which we will return in detail in chapter 6 on homeostasis and health.

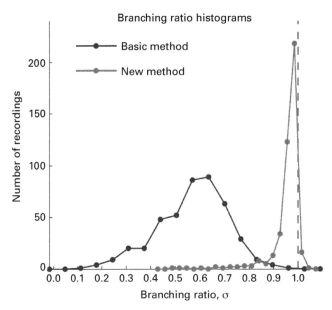

Figure 3.9
The branching ratio is close to 1 in dissociated cultures. Plots are histograms, with number of observations on the y-axis and branching ratio measured on x-axis. The broad curve with the low peak gives results for the basic method of estimating the branching ratio (Beggs and Plenz 2003); it has a median value near σ = 0.65 and is now known to be less accurate. The sharply peaked curve gives results from the new, highly accurate method developed by Wilting and Priesemann (2018). Note that it peaks slightly below the critical value of 1.0. Cultures (*n* = 36) were measured several times each week as they developed, for several weeks. Adapted from Timme et al. (2016b). Data sets freely available at CRCNS data website at https://crcns.org/data-sets/hc/hc-8. Software for estimating branching ratio is freely available in Marshall et al. (2016) and Spitzner et al. (2020).

What should we conclude from the fact that the branching ratio is almost always less than 1? This indicates that the activity of the cortex, in vitro and in vivo, is not exactly at the critical point. It is close to the critical point, but it is not in fact critical. There could be several reasons for this. First, it might be that we need to record from larger numbers of neurons, like thousands, before we will truly know what σ is. This seems unlikely, though, given that the methods of Wilting and Priesemann (2018) are accurate under subsampling. Second, it could be that biology is inherently noisy and that our current measurements are close enough to the value of 1, given experimental error. This may sound plausible, but if it were true then we would expect an equal number of measurements to tell us that σ is slightly greater than 1. Yet we do not see these very often, and we nearly always see values slightly less than 1. Third, this could be an authentic result, telling us that the brain is not critical.

Would the third option undermine the main idea of this book? Not in the least, as being very close to the critical point still brings with it all the benefits of optimal information processing and universality we have been talking about. In fact, the finding that the cortex is not exactly at the critical point was anticipated even before the extremely accurate methods for measuring σ were developed. In 2014, Viola Priesemann and colleagues (Priesemann et al. 2014) compared activity from computational models to empirical data. They concluded that the cortex was operating in a slightly subcritical regime. She and her group have since gone on to argue that this preserves optimal information processing while safely avoiding seizures that could occur if the system crossed over into supercriticality (Wilting and Priesemann 2019b). Also in 2014, we published a theory paper arguing that when a patch of cortex receives inputs from other brain areas, it is driven to a slightly subcritical regime, while still preserving optimal information processing (Williams-Garcia et al. 2014). We called this new regime "quasicriticality," and have since done experiments to test its predictions (Fosque et al. 2021; Helias 2021). There is now much discussion about how the cortex operates in the vicinity of the critical point, and this is an extremely active area of research. We will address this more fully in chapter 7, dedicated to this topic.

For now, we can conclude that the best estimates of the branching ratio consistently show the cortex operates very close to, but slightly below, the critical point. These findings are consistent with the hypothesis that the cortex operates near the critical point to optimize information processing. Let us continue looking in the data for more signatures of being near the critical point.

Data: A Phase Transition with Control and Order Parameters

Clearly, another requirement of a phase transition is the presence of at least two distinct phases. With neural networks, this can be an inactive phase and an active phase. At the border between them, we expect to see power laws. Figure 3.10 shows avalanche size distributions from cortical slice cultures placed on a 512-electrode array. These slices can be

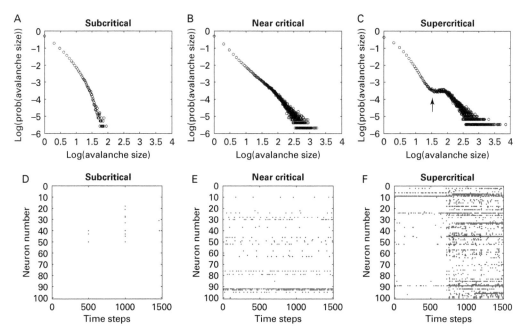

Figure 3.10
Inactive and active phases, with a power law at the transition. *A–C*, Avalanche size distributions from cortical slice networks that are subcritical, nearly critical, and supercritical. These curve down, are straight, and curve upward, respectively. For supercritical case, note dip at arrow. To the left, the distribution curves downward, indicating subcritical avalanches that failed to amplify; to the right it curves upward indicating supercritical avalanches that are larger than in the critical case. *D–F*, Corresponding raster plots during the first 1,500 time steps show transition from very sparse to moderate, then heavy activity in supercritical case.

bathed in solutions that enhance (e.g., muscimol) or reduce (e.g., bicuculline) inhibition to produce activity that is subcritical or supercritical. Often, the slices will show natural variability as well so that some are slightly supercritical or subcritical without pharmacological manipulations. On average, though, untreated slices are near the critical point (Friedman et al. 2012). A similar continuum in avalanche size distributions, from downwardly curved, to straight, to upwardly curved, have been seen from spikes obtained from rat somatosensory cortex in vivo (Gautam et al. 2015). Together, these studies clearly show that there are two distinct phases in cortical networks and that pharmacological manipulations that change the level of inhibition can tune the networks from one phase to the other.

Data: An Exponent Relation between Multiple Power Laws

If networks of living neurons operate near a critical point, they also should produce data that satisfy the exponent relation. To accomplish this, recall that three power laws are needed: (1) the avalanche size distribution, (2) the avalanche duration distribution, (3) the average avalanche size, for a given duration, plotted against the duration. The exponents for these power laws are τ, α and γ, respectively, and they should fulfill the equation $(\alpha - 1)/(\tau - 1) = \gamma$. This prediction was first tested in cortical tissue by Friedman et al. (2012), where we used a 512-electrode array to record spontaneous activity from cortical slice cultures. In figure 3.11 you can see the distributions produced by one such culture. In this example, they fulfill the exponent relation almost exactly.

While not all cultures satisfy the relation this well, most of them are fairly close, as can be seen from figure 3.12. There, 13 cultures are shown to lie on the line with slope $\gamma = 1.31 \pm 0.04$; each of them satisfies the exponent relation within experimental error. Other groups have reported similar results in spike data from turtle cortex (Shew et al. 2015), rat visual cortex in vivo (Ma et al. 2019) and in the cortices of freely moving mice (Fontenele et al. 2019).

Data: Fractal Copies of Avalanches

Recall that yet another prediction of the criticality hypothesis is that neuronal avalanches will have shapes that are rescaled, or fractal, copies of each other. This similarity allows them to be collapsed on top of each other into the same shape, something called shape collapse. In figure 3.13, we see avalanche shape collapse performed for data recorded from a cortical slice culture. Since this was first reported in vitro (Friedman et al. 2012), it has also been done with spiking data from rats in vivo (Ma et al. 2019) and with LFP data from monkeys in vivo (Miller, Yu, and Plenz 2019).

To summarize, spike data from microelectrode arrays indicate that networks of cortical neurons operate close to the critical point. We saw this from six features: (1) A branching ratio very close to 1; (2) Distinct phases of inactivity and activity; (3) The networks can be tuned to one phase or the other by pharmacological manipulations; (4) Power laws are found at the border between the inactive phase and the active phase; (5) An exponent relation among these power laws appears in the phase transition region; and (6) Avalanche shapes are scaled copies of each other. These findings are in good agreement with predictions made by a simple branching model that can be tuned to operate near the critical point.

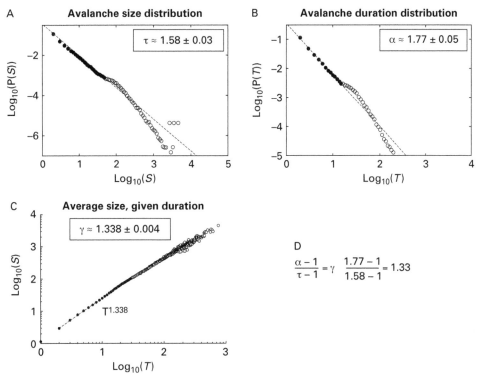

Figure 3.11
Satisfying the exponent relation in a cortical slice culture. *A*, Avalanche size (*S*) distribution follows a power law (filled circles) with an exponential cutoff (open circles). *B*, Avalanche duration (*T*) distribution also follows a power law with an exponential cutoff. *C*, Average avalanche size, for a given duration, plotted against duration. Here, filled circles indicate the portion of the plot that was used to fit the power law. The slight bump at the end of the linear portion of the distributions in A and B is expected and is due to an exponential cutoff caused by the finite size of the number of neurons sampled (Fosque et al. 2021). *D*, The exponents satisfy the exponent relation well. Organotypic cultures were prepared from coronal slices of mouse somatosensory cortex. Data have been previously published in Friedman et al. (2012) and are freely available at https://crcns.org/data-sets/ssc/ssc-3.

Even though it is simple, the branching model still captures many relevant features of the phase transition, and it successfully predicted these features in physiological data from living cortical networks. As we will see later, these predictions have been verified in other preparations as well, including in vivo recordings from awake and behaving animals.

Objections to These Signatures of Criticality

Despite these results, some are skeptical these signatures of criticality are conclusive. There are also questions raised about how we should interpret data that display these signatures. Here, for the sake of completeness, I will mention the nature of these issues and address them.

The first type of objection is that many distributions from neuronal data, when subjected to rigorous statistical tests, are not found to be power laws. The field has responded to this by more carefully analyzing the data. When appropriate methods are correctly applied to both spike and local field potential data, several groups found distributions produced by neuronal avalanches to be significantly better fit by power laws than other

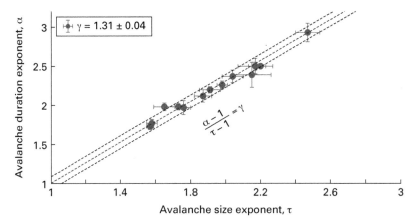

Figure 3.12
Multiple cortical slice networks satisfy the exponent relation. Each circle represents a pair of exponents (τ, α) obtained from a single culture, with error bars showing one standard deviation. Although the exponent pairs vary, they all lie close to a line that is given by the exponent relation, shown in the equation just below the line. The central dashed line is drawn with the average value of γ for these data points. The flanking dashed lines indicate one standard deviation. This shows that all these cultures satisfy the exponent relation, within experimental error. Each experiment was a 1-hour recording from a 512-electrode array. Average number of neurons was 310 ± 20. A plot of this type, with pairs of exponents satisfying the exponent relation, was first shown by Shew et al. (2015) for experiments in turtle cortex.

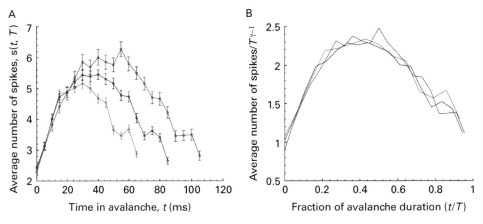

Figure 3.13
Avalanche shape collapse from spiking data. Data collected from organotypic cultures in vitro recorded with a 512-electrode array. *A*, Average shapes for avalanches of durations 65 ms, 85 ms and 105 ms. Each shape is the average of thousands of avalanches. Error bars indicate standard deviations. *B*, Rescaling of the three shapes shows that they collapse with relatively little error onto a single parabola-like shape. Rescaling in time is accomplished by dividing each shape by its duration, making them all of length 1. Rescaling in height is accomplished by dividing the average number of spiking neurons at each time step by the duration of the avalanche (T) raised to the exponent $\gamma - 1$. Recall γ is the exponent from the plot of avalanche size against duration (figure 3.11C). The ability to perform this collapse with the exponent γ is a strong indicator that the network is operating close to the critical point. Modified from Friedman et al. (2012).

competing distributions (Yu et al. 2014; Bellay et al. 2015; Shew et al. 2015; Timme et al. 2016; Ponce-Alvarez et al. 2018). There are now free software toolboxes for performing these tests (Alstott, Bullmore, and Plenz 2014; Marshall et al. 2016), and they are routinely used. To be clear, however, not all groups report that their data are always power-law distributed; some have stated that this depends on the cognitive state of the animal (Hahn et al. 2017) or the type of data or how it is binned (Priesemann et al. 2014). Some deviations from power-law form are understandable and even predicted by the criticality hypothesis. But the fact that power-law distributions can be statistically verified and are found when they are expected has made this objection increasingly rare.

The second type of objection is that many noncritical processes can create power laws. Therefore, the argument goes, a power law is insufficient to demonstrate a phase transition. To illustrate, an ensemble of exponential decay processes with randomly chosen exponents can sum to create a distribution that has a long tail that looks like a power law. This type of situation could occur quite easily in the brain, where it is known that there are very rapid and very long duration decay processes (Fusi, Drew, and Abbott 2005; Gal et al. 2010). For example, the decay of a postsynaptic current in a neuronal membrane is only a few milliseconds, but the turnover of proteins in a synapse can take days to years.

The main response to this objection is to show that power laws caused by noncritical processes do not possess a phase transition. They will be resistant to attempts to change their distribution and will not show a continuum of curves like those we saw in figure 3.10. Conversely, if you can demonstrate a phase transition with a system, then it potentially has a critical point. And if you can show it has multiple power laws that satisfy an exponent relation near the phase transition, then you have excellent evidence for a critical point (Beggs and Timme 2012). It is increasingly common for investigators to look for multiple power laws and to test the exponent relation (Friedman et al. 2012; Arviv, Goldstein, and Shriki 2015; Shew et al. 2015; Ponce-Alvarez et al. 2018; Miller, Yu, and Plenz 2019). Here again, the field has responded in a healthy way to objections.

The third issue is relatively new, namely, that extremely simple models can produce the signatures of criticality we discussed. For example, Destexhe and Touboul (2021) have shown that an Ornstein-Uhlenbeck process, which is like a random walk combined with a restoring force, can satisfy the exponent relation and avalanche shape collapse. The argument is that the signatures of criticality we see in the data may not arise through the collective interactions of neurons. Instead, they could be caused simply by noise and a restoring force. This is a very important question and deserves an adequate response.

To explain this response, it is important to introduce the concept of *degrees of freedom* of a system. When describing any system, a defining feature is the number of relevant variables that must be used to specify it. Systems like a network of neurons, where all the connections must be given, have many degrees of freedom. In contrast, systems like the Ornstein-Uhlenbeck process proposed above have only one degree of freedom: the ratio of noise to the restoring force. Power-law behavior can be observed in systems with many or few degrees of freedom, but the mechanisms leading to those power laws may differ. In general, systems with many degrees of freedom are studied using the approach of *statistical mechanics*, while those with few degrees of freedom are studied with the tools of *dynamical systems*.

In statistical mechanics, criticality emerges through the collective interactions of many similar units. It is manifested by the statistics of the ensemble of units. We have discussed how this can apply to networks of neurons or flocks of starlings as they send signals to each other. Statistical criticality is also applicable to water when it is poised at the phase transition between gas and liquid and describes a piece of iron just on the verge of being magnetized. In all these cases, the essential ingredient is the collective interactions. If those interactions are cut, the signatures of criticality, that is, power-law behavior, disappear and there are no emergent phenomena. The neurons must send synaptic signals to each other; the starlings must sense their neighbors turning; water molecules must collide with each other; and magnetic spins must flip their nearest neighbors. These systems are tuned by changing the relative strength of the interactions between their constituent units.

In dynamical systems, in contrast, signatures of criticality do not require many interacting units. In fact, they can be produced by a single unit interacting with itself. They are manifested by the dynamics of this unit over time. For example, it has long been known that a fair coin flip can show signatures of criticality. If we start at the origin and take one step forward for every head (+1) and one step backward for every tail (−1), after millions of flips we see a random walk that typically returns to the origin many times. Interestingly, the number of flips between these returns will follow a power-law distribution (Kostinski and Amir 2016). Moreover, a reflected coin flip (that is just the absolute value of the excursions) will produce avalanche shapes that can be collapsed onto themselves and multiple power laws that satisfy the exponent relation. But if the dynamics are tuned to be biased so that the coin produces slightly more heads than tails, we will not observe power laws or shape collapse because it will not return to the origin as often. This is analogous to a diminished restoring force. In addition, a system with only 1 unit obviously cannot produce complex emergent phenomena like spiral waves or repeating activity patterns. Nevertheless, we have a very simple tunable model that produces all the signatures of criticality we are looking for.

Now the question is, which type of criticality best approximates the cortex? The signatures by themselves will not be enough to tell us which. But the mechanisms producing these signatures can. Do the power laws and shape collapse emerge as a result of collective interactions as the criticality hypothesis predicts? Or are the signatures best explained by many disconnected neurons being driven by a random noise source? The distinction lies in whether these signatures depend on the neurons' ability to interact with each other.

Experiments give an unambiguous answer to this; they consistently show that critical signatures in neuronal systems depend on communication between neurons. For example, we (Beggs and Plenz 2003) showed that application of picrotoxin, a gamma-aminobutyric acid (GABA) antagonist that blocks inhibitory synaptic transmission, causes disruption of power laws in acute cortical slices. When picrotoxin was washed out, the activity returned toward a power-law distribution. Woodrow Shew, Dietmar Plenz, and collaborators (Shew et al. 2009) showed that application of (2R)-amino-5-phosphonovaleric acid (AP5) and DNQX, blocking excitatory synaptic transmission, disrupted power laws in organotypic cortical cultures. Similar work has been done in vivo with rats, where application of the $GABA_A$ antagonist bicuculline and the $GABA_A$ agonist muscimol have been used to tune cortical activity away from a critical point and into the supercritical and subcritical regimes, respectively (Gautam et al. 2015). These manipulations are not confined to chemical synapses, though. In zebrafish larva in vivo, Ponce-Alvarez and colleagues showed that even the gap-junction blocker heptanol could disrupt the quality of avalanche shape collapse (Ponce-Alvarez et al. 2018), again moving the neural network away from a critical point.

This issue has highlighted the fact that operating near a critical point in living neural networks is better explained by collective interactions than by the dynamics of a single unit. That said, there are cases where an ensemble of neurons can show dynamics that are either damped, critical, or supercritical. Statistical criticality and dynamical systems criticality can sometimes overlap as we will see in the next chapter. The main utility of simple models like the coin flip is that they force us to think about what type of criticality we observe in the data. In this sense, the comments of Destexhe and Touboul (2021) are not really objections to signatures of criticality but rather clarifications about how we should interpret data.

In summary, the issues raised about signatures of criticality are motivated by reasonable questions and are posed by thoughtful scientists. Far from hindering the progress of this field, they have strengthened it by deepening our understanding of exactly what we mean when we claim that a network of neurons is nearly critical. Such discussion should not be suppressed but welcomed as opportunities to learn more about the truths of nature.

Chapter Summary

The criticality hypothesis claims that by operating near a phase transition point, living neural networks optimize information processing. The first component of this hypothesis is concerned with the question of whether cortical networks are in fact nearly critical. To investigate this, we can look to predictions made by a model and then to data to see if these predictions are upheld. The branching model is known to have an active phase and an inactive phase, separated by a critical point. Near the critical point, it produces several power-law distributions, and the exponents of these satisfy an exponent relation. Moreover, at the critical point, both small and large avalanches have profiles that are fractal copies of each other. All these signatures of being near the critical point can be examined in spike recordings from networks of cortical neurons. We find that these data have a branching ratio near 1, a clear phase transition, multiple power laws that satisfy the exponent relation, and avalanches that have the same rescaled shapes. This type of criticality emerges through collective interactions among neurons. Taken together, these results strongly support the hypothesis that these networks operate near the critical point. While there are still some objections to this conclusion, the field has responded to these in a thoughtful manner. Dialogue is likely to continue in this area and should be welcomed.

The seventh signature of criticality, as we mentioned earlier, consists of information processing functions that are sharply peaked at the critical point. This will be the substance of our next chapter.

Exercises for this chapter can be found through a link given in the appendix.

4

Optimality

It doesn't matter how beautiful your theory is, it doesn't matter how smart you are. If it doesn't agree with [the] experiment, it's wrong. *In that simple statement is the key to science.*
—Richard P. Feynman

On several occasions we have had Nobel laureate physicists visit our department to give lectures. It is always exciting to meet them in private over dinner or at tea to ask them questions. Often, they ask us what we are researching. One time, when I gave my three-minute pitch that "the neocortex operates near a second-order phase transition to optimize information processing," this particular laureate replied after a few moments by saying: "Of course! It has to be like that. I am surprised this has not already been established." While I was comforted by his response, I also realized that he was removed from the particulars of the data and the sometimes heated discussions surrounding it.

While it is fairly easy to state that the cortex should optimize information processing, it is more difficult to break down the specifics of what this would mean. The cortex is probably the most universal of all computational tissues in the brain. Its rapid expansion, in evolutionary terms, made it inherit a broad range of functions from sensory processing to motor commands, as well as associations and executive processes. Given this, it is fascinating that a patch of cortex has a remarkably similar structure no matter where it is located or what functions it is thought to perform. This common cortical structure consists of six layers of neurons arranged in what has been called a canonical circuit (Douglas, Martin, and Whitteridge 1989). The function of this circuit seems to be largely determined by the inputs it receives. For example, when auditory cortex is given visual inputs, the newly repurposed patch of cortex adapts and starts to perform some visual functions (Sharma, Angelucci, and Sur 2000).

In addition to this diversity of brain functions, the specific information processing tasks ascribed to the cortex are equally disparate. A given patch of cortex is thought to transmit information, store it, and compute, all while being exquisitely sensitive to changes in inputs over a huge dynamic range. How can one type of circuit be simultaneously good at so many different brain functions and information processing tasks?

The claim of the criticality hypothesis is that the cortex can optimize all these information processing tasks at the same time by operating close to the critical point. This idea is appealing and perhaps even beautiful, but these virtues do not necessarily make it true. Can this idea be scientifically tested, as Richard Feynman would insist?

In this chapter, I would like to get into the details of experiments that directly address this question. As we noted, the nearly scale-free power laws produced close to the critical point allow information across many different scales to mix. Earlier, we described intuitively how this would facilitate information transmission through cortical networks. Here we will investigate this claim by using the branching model as our guide. In addition to looking at information transmission, we will also probe dynamic range and susceptibility, which are related to the ability of the cortex to respond to minute changes in inputs even as they vary widely in magnitude. Prompted by these predictions of the branching model, we will then review experiments with cortical tissue that have tested them. After that, we will briefly describe other predictions of the branching model that have not yet been tested. Many questions remain open, leaving plenty of work to be done in the future.

Before we go on, I want to emphasize that the information processing functions we will examine in this chapter all emerge through the collective interactions of neurons. For example, susceptibility for a network of neurons can diverge to infinity at the critical point, something that would be impossible for an individual neuron. Thus, the network has properties that single neurons by themselves cannot have. We will be exploring emergent information processing whose power is maximized near the critical point.

The Branching Model: Information Transmission

To address information transmission, we will revisit the branching model network configured as a series of layers of neurons, each receiving inputs from the previous layer and projecting to the subsequent layer (figure 4.1A). The results that we obtain here also apply to recurrent networks, but it is easier to demonstrate in a layered feed-forward network. As before, we will present a configuration of active neurons to the first layer, let it propagate through all the layers, and then read out the configuration of active neurons in the last layer. We will measure the mutual information between the input and the output as a function of the branching ratio (figure 4.1B).

As we can see from the figure, there is a peak at a branching ratio of $\sigma = 1.3$, which is near the critical value of 1, but not exactly there. In a later section, I will explain why the peak differs from what we expected. For now, let me just say it is related to the size of the network simulation. For a much larger simulation, it would peak at exactly 1. To understand why the mutual information has a peak at all, it will be necessary to explain how information transmission through the network is calculated.

Understanding Mutual Information

All measurements of information can be framed in terms of a reduction of uncertainty. Let us consider a simple example to illustrate. Suppose I had a square box partitioned into four equally sized compartments. I place a ball in one of the quadrants (figure 4.2A), and you are to guess where the ball is by asking only simple yes or no questions, as few as possible. Your first question might be: Is the ball in the north half of the box? I answer: No. You follow with: Is the ball in the east side of the box? I answer: Yes. Let us analyze this successive reduction of uncertainty using information theory.

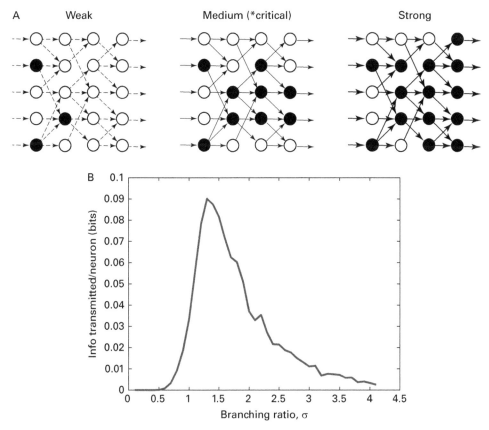

Figure 4.1
Information transmission peaks near the critical point. *A*, Schematic shows three feed-forward networks with the same two neurons activated at the input layer. For the subcritical network, activity dies out quickly; for the critical network the number of active neurons is preserved in the output layer; for the supercritical network the entire output layer is activated. *B*, Mutual information between the input and output is plotted as a function of the branching ratio, the control parameter. Note the steep rise with a peak at a branching ratio of 1.3. The feed-forward network had 49 neurons in each layer and 15 layers; curve shows average values after 16 runs.

At first, the ball would seem to have an equal probability of being in any of the four quadrants. We can represent this by a distribution with a probability of 0.25 for each of the quadrants. This distribution is flat. After the answer to your first question, you know the ball must be in one of the two southern quadrants. Since there are only two, each has a probability of 0.5 for containing the ball. Now the distribution is no longer flat but has two bins with zero probability and two with 0.5. Your final question distinguishes between the eastern and western halves of the remaining area. Now you know the ball is in the southwest quadrant. This produces a distribution that is peaked, with a value of 1.0 for the southwest quadrant and zero for the other three quadrants (figure 4.2C).

Each time you asked a question, you cut the space of possible locations for the ball in half. And each time you were able to reduce the uncertainty by one half, you gained information. Let us now try to quantify this. The uncertainty at each stage can be measured by the entropy of the corresponding distribution. The entropy of a distribution, symbolized by *H*, is given by this equation:

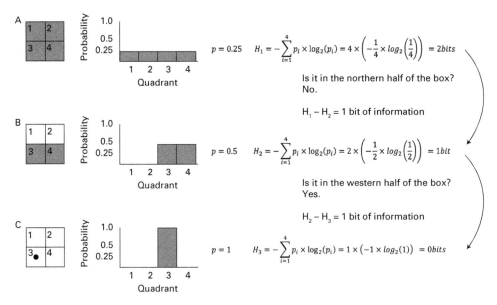

Figure 4.2
Information explained. *A*, A ball is hidden in one of four quadrants in a box. Each quadrant is equally likely to contain the ball, so the distribution has four bars, each with probability of 0.25. When the entropy is calculated, it comes to 2 bits. This is the uncertainty of the ball's location. *B*, After an answer to a binary question rules out the northern half of the box, the uncertainty is reduced by half. The distribution now has two bars, each with probability of 0.5. The entropy is 1 bit. Notice that the uncertainty was reduced from 2 bits to 1 bit. This reduction in uncertainty is information, amounting to 1 bit. *C*, After the answer to another binary question rules out the eastern side of the box, the uncertainty is reduced again by 1 bit. The entropy of the final distribution is zero. The answer to each question provided 1 bit of information because it reduced the uncertainty by half.

$$H = -\sum_{i=1}^{N} p_i \times log_2(p_i).$$

Here, each p_i value is the probability of finding the ball in the ith quadrant. Notice that the logarithm is in base 2. This is a convenient base because it allows us to measure the results of our binary choices (north versus south, east versus west). Every time you cut the space of possible locations in half you reduced the uncertainty by a power of two ($1/2 = 2^{-1}$). Each power of two is a binary unit, which is called a bit. A logarithm with base 2 measures these powers, or bits, easily ($-log_2(1/2) = 1$ bit). Applying this measure to the first distribution, where all the quadrants have a probability of 0.25, we have $H_1 = 2$ bits. For the second distribution, we have $H_2 = 1$ bit. And for the third distribution, we have $H_3 = 0$ bits. Using this measure, we can see that each yes or no binary question reduced the uncertainty by 1 bit. At the end, there was no uncertainty at all. A bit is then the information that is gained by halving the uncertainty.

Let us take this framework and apply it to the problem of quantifying information transmission through a network. To do this, we measure the mutual information between the input and the output.[1] The mutual information, *MI(In;Out)*, is given by this equation:

$$MI(In;Out) = H(Out) - H(Out \mid In).$$

Here, $H(Out)$, the entropy of the output, measures the uncertainty in all the different outputs produced by the network. If there is a wide variety of output configurations seen, each equally often, this will be large. If there is only one type of output ever given, this will be zero. Let us get an intuitive feel for how the diversity of outputs affects the amount of information that can be transmitted. Suppose you are asked to rate a movie and you are given only two choices: (Like, Dislike). If we compare this limited repertoire to four choices (Excellent, Good, Fair, Poor), it is clear that more responses mean that more information can be sent. For our network, note that responses can differ not only in the number of neurons that are active, but also in their arrangement. For example, if we have two neurons in the output, they could have four configurations $(00, 01, 10, 11)$, which is more than the number of neurons that could be on $(0, 1, 2)$. This emphasizes that $H(Out)$ measures not only the number of neurons, but also their arrangement.

How then should we interpret the second term, $H(Out \mid In)$? This entropy measures the uncertainty of the output for a given input and is sometimes called the "equivocation." For mutual information to be high, it is important for the network to have very repeatable responses every time the same input is presented. To illustrate, suppose we presented this input configuration to the network: (01). A network that always responded with an output of (11) would have a low equivocation. In contrast, a network that responded with (11) and (01) equally often would have more equivocation and would be less reliable.

Mutual information between the input and the output of a network depends on two factors then. First, there must be a wide variety of different outputs. This produces a high output entropy. Second, there must be a reliable pairing between a given input and a given output. This produces a low conditional entropy. The difference between these entropies is the mutual information between the input and the output: $MI(In;Out)$.

Understanding Why Information Transmission Peaks near the Critical Point

Given what we now know about how mutual information is measured, we can explain why there is a peak in the curve near the critical point. When the branching ratio is small and the network is subcritical, activity is damped and there is low output variability—everything is off. When the branching ratio is large and the network is supercritical, activity is amplified and there is again low output variability—everything is on. From the perspective of the first term in the mutual information equation then, $H(Out)$, it is clear that being near the critical point produces the highest output variability.

But as for the equivocation, $H(Out \mid In)$, things are only slightly different. For low branching ratios, this term is low because no matter what input, the output tends to be the same (most neurons off). And for high branching ratios, this term is again low because no matter what input, the output tends to be the same (most neurons on). As you might expect, this term is highest for branching ratios near the critical point.

If both $H(Out)$ and $H(Out \mid In)$ rise near the critical branching ratio, then how can information transmission be maximized there? While both rise, $H(Out)$ rises a little bit more than $H(Out \mid In)$ near the critical point, causing the difference $H(Out)—H(Out \mid In)$ to be maximized there. It turns out the reason the mutual information is greater is largely because the output variability is bigger near the critical point. This understanding will help us in explaining why dynamic range is also maximized near the critical point, as we will discuss next.

The Branching Model: Dynamic Range

Our sensory systems are tasked with representing a very wide range of input strengths. Consider, for example, the differences in light levels between a dark room and a white sandy beach on a sunny day. Here, the range in luminance values may span nearly eight orders of magnitude. While typical daily differences are not that extreme, our visual system routinely processes light levels spanning four orders of magnitude (Radonjić et al. 2011). Equally impressive, our auditory system can discern sound intensities differing by 10 orders of magnitude or more (Berg and Stork 1995). Fortunately, the brain does not represent this huge range of numbers by employing a correspondingly huge range of neurons; there is massive compression in its representation. Still, the range of its representation must be as broad as possible.

For information about widely varying stimulus strengths to be represented with minimal loss of information, it is important for the entropy of the output distribution of a network, $H(Out)$, to be large. This can be captured by measuring the dynamic range of the outputs that represent the stimuli. To begin to quantify this, let us measure the output strength as just the number of neurons on at the output layer, ignoring the many different configurations that those output neurons could have. In this reduced format, to measure the breadth of the output distribution, we will consider the upper 80th percentile and the lower 20th percentile of the response distribution. The difference between the number of neurons that are on in these two conditions can serve as our proxy for the dynamic range.

In figure 4.3 we see the output distributions for three branching model networks that had three layers and 64 neurons in each layer. When the branching ratio is low, the number of neurons on in the output layer has a relatively narrow peak around 10. This should not be surprising, as activity is damped in the subcritical case. When the branching ratio is high, there is a narrow peak around 60, nearly saturating the network as we would expect in the supercritical case. When the network is nearly critical, the output distribution is at its broadest, with anywhere from 23 to 43 neurons on in the middle 60 percent of the distribution, giving the largest dynamic range. Using a network model with a tunable

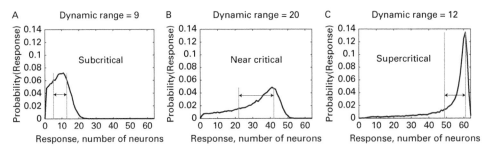

Figure 4.3
Dynamic range is maximal near the critical point. A feed-forward network with 64 neurons per layer had 1 to 64 neurons sequentially activated at the input layer. The output was the number of neurons active in layer 3. *A*, When the network is subcritical, very few neurons are activated in layer 3, even for maximal stimulation. The output had a relatively narrow distribution around 10. Vertical lines indicate the 20th and 80th percentile; the dynamic range is the distance between these lines (arrow) and is 9. *B*, When the network is near the critical point, the response distribution is broader, and has a dynamic range of 20. *C*, When the network is supercritical, almost all neurons in layer 3 are activated. The response distribution has a narrow peak around 60 with a dynamic range of 12.

branching ratio, Kinouchi and Copelli (2006) predicted that dynamic range would be maximized at the critical point for neural networks. As we will see later, these important results motivated experimental tests of this idea.

The Branching Model: Finite Size Effects

Now we should return to the question of why the information transmission curve in figure 4.1B does not peak at a branching ratio of exactly 1. As I mentioned before, the reason behind this has to do with the size of the model we are using. When we obtain information transmission curves for smaller models, as shown in figure 4.4A, we notice two trends. First, the peaks get taller and narrower as the models get larger. Second, the peaks move closer to a branching ratio of 1 as the models get larger (figure 4.4B). This suggests that in the limit of an infinitely large model, the curve would be extremely tall and narrow, centered at a branching ratio of exactly 1.

When we talk about a phase transition or a critical point in the sense of statistical physics, we are considering what would happen in the limit of infinite model size. This is also called the "thermodynamic limit." In our model networks, however, we have only finite sizes, and sometimes rather small sizes at that. For reasonable computation times, all the simulated networks in this chapter are relatively small, with a few tens to hundreds of neurons at most. The fact that these small models do not reproduce the same results that much larger models would can be attributed to what are called "finite size effects." Yet even with these finite sizes, we can still get an idea about what the infinite limit would be by looking at how our results scale with model size. For example, the plot in figure 4.4B suggests that for very large model sizes, the peak would approach a branching ratio of 1. A proper analysis of this would take much computer time to perform with substantially larger models, containing perhaps tens of thousands of neurons. But even in our very limited range of models here, we can see that this type of trend is plausible. In Beggs and

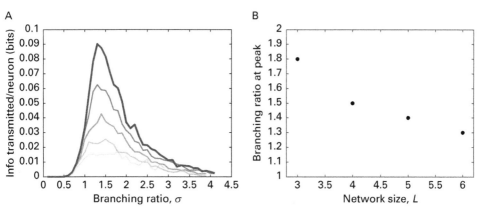

Figure 4.4
Peak of information transmission depends on network size. *A*, The information transmitted per neuron through feed-forward networks, as the branching ratio is increased. Multiple network sizes are shown. Lowest curve (lightest gray) is a 3×3 network with 9 units in each layer; highest curve (darkest gray) is a 7×7 network with 49 units in each layer. Note how curves get taller as network size is increased, and how peaks move to the left. *B*, Branching ratio at which information transmission peaks gradually moves toward 1 as network size, *L* (3,4,5,6) is increased. This suggests that for a network of infinite size, it would peak at exactly 1.

Plenz (2003), for example, a curve was fit to a simple branching model network with four layers and 6 to 14 neurons per layer; it was found to asymptote at a branching ratio of 1.04 ± 0.10.

These finite size effects will impact the interpretation of experimental results. Great care should always be taken to extrapolate toward the thermodynamic limit to get a proper estimate of how close to the critical point an actual network may be. Notice that this also suggests results from a relatively small sample of neurons should not be expected to point exactly toward the critical point. Rather, some results will appear to be subcritical or supercritical, depending on the sampled population size and the type of measurement being taken.

One more point needs to be made about these narrow, peaked curves. As we have seen, several quantities that we can measure in these networks will show a peak near the critical point. In the thermodynamic limit, these curves would become infinitely tall and narrow. If we can again use terminology from statistical physics, such curves are said to "diverge." These divergences are a hallmark of the critical point, but in small simulations or in small samples of data, we should only expect curves that show a peak.

The Branching Model: Susceptibility

In addition to dynamic range, another important feature of sensory systems is that they be able to detect slight changes in inputs. For example, a sensory network might be receiving a stimulus that drives 5,000 of its 10,000 receptor neurons—would it be able to sense if the number of active receptor neurons changed to 5,001 or 4,999? Notice that this is not necessarily the same thing as being able to represent a wide range of input strengths, as we discussed for the dynamic range. It is possible that a network could have a very narrow dynamic range, and yet be highly sensitive to changes within that range.

This sensitivity to changes in inputs can be captured by the susceptibility, Chi (χ), from statistical physics. The equation for it is:

$$\chi = \langle \rho^2 \rangle - \langle \rho \rangle^2,$$

where ρ is the density of active sites and the angled brackets indicate an average over time. In words, the susceptibility is just the variance over time of the number of active neurons.

For subcritical networks, the number of active neurons is usually very low, so the variance is also low. Activating an additional neuron may cause an uptick in activity, but this will be quickly damped. In contrast, for supercritical networks, the number of active neurons is typically very high, with only slight changes over time, again leading to a low variance. Turning an additional neuron on or off in supercritical networks will make very little difference in the overall population activity because they are nearly saturated already. In networks near the critical point, activity levels can vary substantially, and turning one neuron on or off can cause widely propagating effects.

In figure 4.5, the density of active sites and the susceptibility in a branching network are plotted for different values of the branching ratio. The density of active sites shows a jump from zero to increasing values at the critical branching ratio of 1, following the pattern we saw before in figure 3.3A. On the same axes, we see the corresponding value of the susceptibility peak near this phase transition point, at a branching ratio of 1.6. Recall that finite size effects will cause it to not peak at a branching ratio of exactly 1. This model

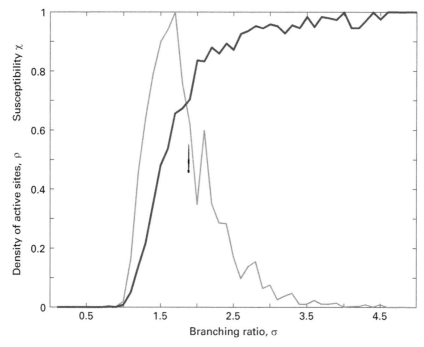

Figure 4.5
Susceptibility peaks near the critical point. The susceptibility, χ, is plotted in gray against the branching ratio, σ, for a feed-forward branching network with 50 neurons per layer and 25 layers. For comparison, the density of active sites, ρ, is plotted in black. The susceptibility curve peaks roughly in the transition region between the inactive and active phases. Finite size effects prevent it from peaking at a branching ratio of exactly 1.

predicts that neural networks in the brain would be able to detect small changes in stimuli best when they operate near the critical point.

Now that we have seen the predictions of the branching model for information transmission, dynamic range, and susceptibility, we can turn to review the neurophysiological experiments that were done to test these ideas. Each experiment has many details that must be understood before the results can be properly interpreted; this will be the substance of the next few sections.

Data: Dynamic Range

Although we demonstrated in figure 4.3 that the branching model has its largest dynamic range when it is tuned to the critical point, this prediction was first made by Osame Kinouchi and Mauro Copelli (Kinouchi and Copelli 2006), using a branching model (figure 4.6). This prediction was tested experimentally by Woodrow Shew and colleagues (Shew et al. 2009), forming the second in a pair of elegant papers.

To perform these experiments, Woodrow Shew, Dietmar Plenz and colleagues grew organotypic cultures on 60-electrode arrays where they could record LFPs (local field potentials) and stimulate with current pulses (Shew et al. 2009). They injected current through an electrode near the center of the array, placed in superficial layers 2 and 3 of cortex, and were able to vary the stimulus intensity from 6–200 μA. In response to these

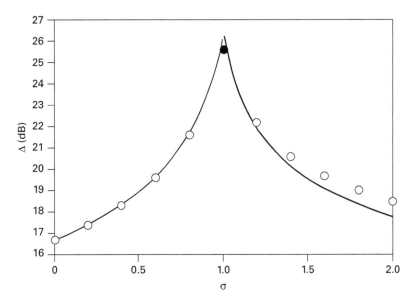

Figure 4.6
Dynamic range predicted to be maximal at $\sigma = 1$. Dynamic range ($\Delta(dB)$) along y-axis, branching ratio σ along x-axis. Model results given as circles; theoretical predictions given by curved lines. When the model is tuned to the critical point, dynamic range is greatest. Adapted from Kinouchi and Copelli (2006).

currents, the network produced LFPs on various numbers of electrodes. As you might expect, larger currents generally produced more electrodes with suprathreshold LFPs. The factor of interest here, though, was not just the number of LFPs, but how well the range of those numbers corresponded to the range of stimulus strengths.

We will now describe how they tuned the networks to be subcritical and supercritical, as well as how they measured these different regimes. As we mentioned before, unperturbed cultures tend to operate near the critical point; pharmacological manipulations are usually needed to push them into subcritical or supercritical states. To produce subcritical activity, they applied agents that blocked excitatory synaptic transmission (DNQX and AP5). For supercritical activity, they applied picrotoxin (PTX), which blocked fast inhibitory synaptic transmission.

To assess the state of the network, Shew and colleagues developed a novel measure they called kappa (κ) (Shew et al. 2009). This measure was based on the assumption that a culture near the critical point would produce an avalanche size distribution that followed a power law with a slope of -1.5 in a log-log plot, as was seen in Beggs and Plenz (2003). Subcritical networks produce a downwardly curving distribution, falling below the straight line of this power law, while supercritical networks produce a distribution that curves upward at the tail, rising above the power law, as we saw in figure 3.10C. To quantify these differences, they subtracted the actual distribution from the critical power-law distribution. When networks were subcritical, this showed negative differences; when they were supercritical, this showed positive differences. And when networks were near the critical point, differences near zero were found. They defined their index, κ, as just 1 plus the differences between the distributions. Thus, $\kappa \approx 1$ identified networks close to the critical point, while $\kappa < 1$ corresponded to subcritical networks and $\kappa > 1$ to supercritical networks (figure 4.7).

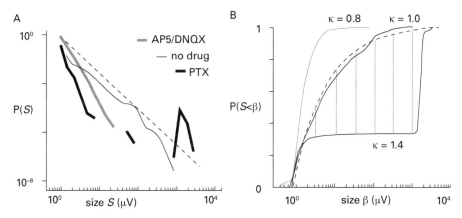

Figure 4.7
The kappa measure for assessing the state of the network. *A*, Avalanche size distributions for networks treated with excitatory transmission blockers AP5/DNQX, no drug, or inhibitory transmission blocker PTX. The dashed line gives the slope of −1.5 expected for a critical branching model with all-to-all connections. Note that the untreated network follows relatively close to the dashed line, while the network treated with AP5/ DNQX points downward and the network treated with PTX curves downward and then upward with a large bump at the tail on the right. *B*, Cumulative distributions for all three networks plotted along with the ideal critical network, shown in dashed line. The curve for $\kappa = 0.8$ corresponds to the AP5/DNQX case and quickly rises to its maximum value at very low avalanche sizes, measured in total μV. The nearly critical $\kappa = 1.0$ case curves gradually upward, following the dashed curve, showing that nearly all avalanche sizes are represented. The supercritical $\kappa = 1.4$ case (PTX) begins to curve upward, then is nearly horizontal, showing that intermediate avalanche sizes are not represented. At the end, the large bump causes a rapid rise in the cumulative distribution. The measure κ can clearly distinguish subcritical, nearly critical, and supercritical networks. Adapted from Shew et al. (2009).

To validate this measure, they constructed a branching model and showed that the value of κ strongly correlated with the value of the branching ratio, σ, set in the model.

To measure dynamic range, they used the amount of stimulus current delivered as the independent variable, and the total amplitude of all the suprathreshold LFP responses, in microvolts (μV), as the dependent variable. This allowed them to make plots as shown in figure 4.8A for each network. The dynamic range, Δ, was related to the distance, in μV, from the 10th percentile to the 90th percentile of the response distribution.[2] Through natural variability of the networks and through applying pharmacological agents, they obtained a range of κ values.

When they plotted the dynamic range observed, Δ, against the parameter κ, they saw a peaked function whose maximum was near the value $\kappa = 1$ (figure 4.8B). With a larger number of electrodes, we would expect this peak to become sharper and move closer to the critical value of $\kappa = 1$, according to the finite size effects we discussed earlier. Although biology is noisy, this curve is undeniably peaked, and it qualitatively follows the predictions made by the branching model (figure 4.6). For low values of σ, dynamic range is expected to be small because activity does not propagate, making all responses small. For high values of σ, dynamic range will again be small, but now because nearly all responses are very large. The greatest dynamic range is expected near the critical point, when $\sigma \approx 1$.

This experiment was thus an important empirical confirmation of the prediction made several years earlier by Kinouchi and Copelli (2006). This was the first test of a prediction from the hypothesis that networks of cortical neurons optimize information processing

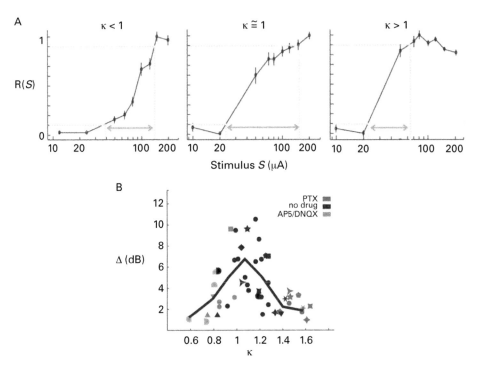

Figure 4.8
Maximal dynamic range near the critical point. *A*, Stimulus-response curves plotted for networks representing the three different regimes determined by κ. Stimulus intensity in μA is given along the x-axis; the cumulative fraction of the response is given along the y-axis. Dynamic range is indicated by the double-headed arrows, which span the range marked out by the 10th to the 90th percentile of the response curve. Note that the broadest arrow is in the panel where κ ≈ 1. *B*, Dynamic range plotted against κ reveals a peaked function. Untreated cultures are represented by black markers. All cultures treated with PTX are in dark gray and occur only in the range of κ > 1.3. All cultures treated with AP5/DNQX are in lighter gray and occur only in the range of κ < 1. Distinct shapes of markers indicate when a single culture was recorded under two conditions. Adapted from Shew et al. (2009).

near the critical point, and it survived refutation. Not only that, but the quantity tested, the dynamic range, had clear functional benefits for an organism's ability to detect and respond to sensory inputs (Shew and Plenz 2013). This experiment therefore also underscored the relevance of criticality to actual information processing tasks that the organism would face every day. This work prompted later in vivo experiments of sensory detection and discrimination under more realistic behavioral conditions, performed by Woodrow Shew and his lab. There, in a whisker deflection task performed in vivo in rats, dynamic range was again maximized when somatosensory cortex was closest to the critical point (Gautam et al. 2015).

Data: Information Transmission

Continuing with testable predictions, we will now turn to examine information transmission through a network. This prediction was first tested by Woodrow Shew, again working with Dietmar Plenz and colleagues, using organotypic cultures of cortex grown on microelectrode arrays (Shew et al. 2011).

Their experimental set up for stimulation was very similar to that described in the pre-vious section on dynamic range. Stimulation with 10 different amounts of current (rang-ing from 10–200 µA) was delivered through a single electrode located near the center of the array, in cortical layers 2 and 3. They repeated each of the 10 different amounts of current 40 times, for a total of 400 stimulations. In response, the network produced LFP activity that could occur on any of the nonstimulated electrodes.

What differs from the dynamic range experiments before is how they considered the network response. They set a threshold of -8 standard deviations for an LFP signal to be considered suprathreshold. If it exceeded this threshold, it was considered active (1). If it did not, it was considered inactive (0). Notice here that the activity on each electrode was treated as a binary variable (1 or 0), and that value of the LFP amplitude in µV was not the variable of interest. Treated this way, the response to any stimulus was the *pattern* of activity across all the electrodes, rather than the total amplitude of the LFPs. To elaborate further, if there were 60 electrodes in the array that could be either on or off, then there were $2^{60} = 1.15 \times 10^{18}$) possible response patterns that could be produced. In practice, far fewer electrodes showed responses, so the total size of the response repertoire was much smaller.

Let us now give some details of how they measured mutual information between the stimulus and response. For stimulation, there were 10 levels of current that could be deliv-ered. For the response, there were 2^{N} possible output patterns, where N is the number of electrodes that could show responses in that experiment. Recall that mutual information is measured by this equation:

$$MI(S;R) = H(R) - H(R \mid S),$$

where we have replaced Input and Output by stimulus (S) and response (R) for the context of this experiment. This states that mutual information will be high when the entropy of the response repertoire is high ($H(R)$ is large) and the equivocation between a given stim-ulus and response is low ($H(R \mid S)$ is small). In simpler words, there should be many dif-ferent response patterns seen as the different currents are delivered. But for a given stimulation current, the same response pattern should be reliably evoked. With this method, they were able to quantify the information transmitted through each culture.

Care must be taken in these types of experiments when small numbers of responses occur. Given that each stimulus current intensity was presented only 40 times, it is possible that the distribution of all responses was subsampled. This could lead to biases in entropy estimation, and an overestimation of the amount of mutual information[3] (Panzeri et al. 2007; Magri et al. 2009). To try to correct for this, Shew and colleagues coarse-grained their array by grouping together 4 electrodes and treating them as one super-electrode (Shew et al. 2011). With this, instead of having an 8×8 array with 64 electrodes (the corner elec-trodes are missing, so there were only 60), they could have a 4×4 array with 16 electrodes, each being a super-electrode with 4 electrodes (again, the missing corner electrodes could lead to four cases where there would be 3, rather than 4, electrodes in a super-electrode). A super-electrode would be considered on if one or more of its electrodes was on. Coarse-graining like this lowers the dimensionality of the responses, somewhat lessening the pos-sibility of undersampling and errors in estimation of mutual information.

As before, they quantified the state of each culture with the parameter κ. In addition to seeing natural variability of κ among the cultures, they could also manipulate κ by applying pharmacological agents as before. When they measured mutual information as a function of κ, they obtained a peaked function, as predicted by the branching model (figure 4.9). This was true when they used data from the entire electrode array, as well as when they used data from the coarse-grained array. Again, the curves would very likely become more sharply peaked if the system size were greater, in accordance with our understanding of finite-size effects.

Although this work may seem in some respects like a replication of the previous experiment on dynamic range, it was a more rigorous test of the predictions that stem from the criticality hypothesis. This is because the mutual information depends on two potentially independent measures, the response entropy and the equivocation, while the dynamic range depends on only one measure, the breadth of the response distribution. Careful readers may note that there is a similarity between the dynamic range and the entropy of the response distribution, though this is not exact. Thus, the experiments that showed information transmission was maximized near the critical point (κ ≈ 1) were a significant advance over what had been done previously. This result again shows that operating near the critical point produces benefits that are expected to have a direct impact on the survival of an organism.

In a continuation of this work, Erik Fagerholm and colleagues, including Woodrow Shew (Fagerholm et al. 2016), looked at information transmission in mouse cortex in vivo.

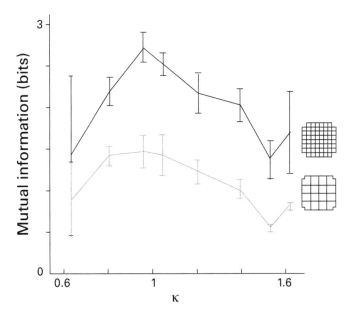

Figure 4.9
Information transmission is maximized at κ = 1. Mutual information between response (R) and stimulus (S), in bits, is given along the y-axis. Values of κ are along the x-axis. Two curves are shown: Black for all 60 electrodes—an 8 × 8 array with corner electrodes missing, as shown in the corresponding grid icon at right. Gray curve shows results when electrodes were grouped into clusters of about 4, so that there were 16 clusters (grid icon at right). Note that corner clusters only contained 3 electrodes. If any of the electrodes in the cluster was supra-threshold, the cluster was considered a 1 and not a 0. Both black and gray curves peak at κ = 1. Error bars indicate standard deviations. Adapted from Shew et al. (2011).

Consistent with the in vitro findings, they showed that information transmission again peaked near the critical point, where the parameter κ was approximately 1.

Data: Susceptibility

The susceptibility quantifies how sensitive a network is to small changes in its inputs. To test the predictions of the branching model, we are interested in seeing how the susceptibility varies with distance from the critical point. One way to assess this would be to manipulate cortical networks to push them away from the critical point. We saw this in the previous work on the dynamic range and information transmission, where pharmacological agents were applied to reduce inhibition or excitation. There is another approach that can be taken that is more natural, though. In long recordings with many neurons, it is possible to observe spontaneous variations in the network state over time. This would be like watching subtle changes in a person's heart rate or blood pressure over an hour.

When a 1-hour recording is broken up into 10-second segments, we can measure the branching ratio in each segment. This is possible because of a newly developed technique for estimating the branching ratio (Wilting and Priesemann 2018). As we can see from figure 4.10A, the branching ratio indeed bounces around over time, but consistently returns to a mean value of $\sigma = 0.9955$, slightly below the critical point. When we measure the susceptibility in each of these segments, we can plot it against the branching ratio, revealing a sharp peak near the mean branching ratio (figure 4.10B). The results from this and other cortical slice cultures show that the predictions of the branching model are again upheld. Networks of cortical neurons will be most sensitive to slight changes in their inputs when they operate near the critical point. Sharply-peaked functions, like what we see with the susceptibility here, are yet another signature of criticality.

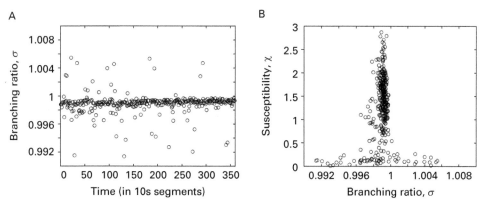

Figure 4.10
Susceptibility peaks near the critical point. *A*, The branching ratio σ naturally varies over a one-hour recording, though it remains very close to the critical value of 1. *B*, When the susceptibility χ is plotted against these branching ratios, it reveals a peak slightly below (~ 0.998) the critical point. Spike data were collected from a cortical slice culture with 233 neurons that was placed on a 512-electrode array. Activity over a period of one hour was broken up into 360 segments, each 10 s long. The susceptibility χ and the branching ratio σ were estimated independently for each segment. Unpublished data from the Beggs lab.

Other Predictions Yet to Be Tested

Before closing this chapter, I would like to briefly show three more examples of functions that are predicted to be optimal near the critical point. For these cases, we do not yet have experimental evidence, as this is still a developing field. These predictions are therefore offered with the hope that they may serve as inspiration for someone to test them experimentally.

Information Storage

Preserving information in memory is another major task performed by cortical networks. There is some experimental evidence that information can be stored for long durations in the patterns of activity from neuronal avalanches. Dietmar Plenz and I examined this by performing 10-hour-long recordings from cortical slice cultures (Beggs and Plenz 2004). We found that some of the spatiotemporal patterns that spontaneously occurred in avalanches would repeat themselves almost exactly many times over the recordings. Our work showed that these avalanche patterns were not random, and that they retained significant information over many hours. Recall the repeating activity patterns of the simple neural network model we discussed in chapter 2 (e.g., figure 2.13).

Motivated by this, Clay Haldeman and I later used the branching model to explore how the number of significantly repeating avalanche patterns varied as the branching ratio was tuned closer to the critical point (Haldeman and Beggs 2005). We found the largest number of these stable, repeating avalanche patterns occurred when the network was near the critical point (figure 4.11). When the network was subcritical, transmission between neurons was too weak to produce many repeating patterns. And when the network was supercritical, the connections were so strong that they often drove the network into only one state where all neurons were on. Thus, a large branching ratio reduced the diversity of stored avalanche patterns. A branching ratio of 1 offered a middle ground where the connections were not too strong and not too weak, producing a multitude of coexisting

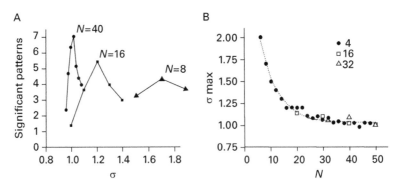

Figure 4.11
Information storage is predicted to be maximal near the critical point. *A*, Number of significantly repeating avalanche patterns (y-axis) plotted against branching ratio, σ, (x-axis). Three curves for increasing model sizes (*N*=8, 16, 40) have peaks that move closer to the critical branching ratio of 1.0. *B*, Branching ratio at which number of repeating patterns is maximized. Fitted curve asymptotically approaches 1.03 ± 0.01 as model size *N* increases, regardless of number of connections per neuron (4, 16, 32). Modified from Haldeman and Beggs (2005). A link to software for generating these plots is in the appendix.

patterns. Our later work showed that the distribution of connection strengths that maximized the number of repeating avalanches was also the distribution that best fit the data (Chen et al. 2010). These predictions have yet to be tested experimentally.

Computation

Moving now to another subject, cortical networks are also essential for computations. Although there is not a single, commonly accepted measure for the computational power of a network, Wolfgang Maas and colleagues have developed a framework that is widely used (Maass, Natschläger, and Markram 2002). To give an overview of this framework, the computational power of a network results from its ability to do two things: (1) separate different inputs by producing different outputs, and (2) group together similar inputs by producing the same outputs. The first of these abilities can be quantified by using something called the network-mediated (NM) separation; for our purposes here, the NM separation can serve as a proxy for computational power.

Nils Bertschinger and Thomas Natschlager (2004) examined how the computational power of a network varied as it was tuned to be near the critical point. Even though they did not use a branching model like the one we have been exploring here, their model shared some similarities and could be tuned by changing the connection weights to each neuron. When the distribution of connection weights had high variance, the dynamics of the model were chaotic; when the distribution had low variance, they were damped and stable. An intermediate variance produced what they called critical dynamics, which varied only slightly for different values of external drive (figure 4.12). With this model, they

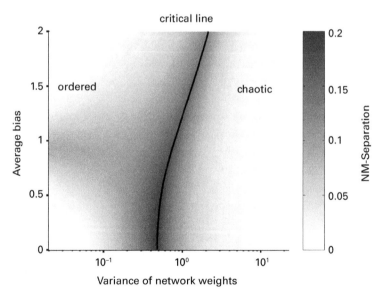

Figure 4.12
Computational power is predicted to be maximal near the critical point. The phase plot of the neural network model used by Bertschinger and Natschlager (2004) shows an ordered dynamical regime (*left*) and a chaotic regime (*right*), separated by a critical line (black). The phase of the network is determined by the variance of the network weights, along the x-axis, and the average amount of external drive (bias) delivered to each model neuron. The grayscale intensity gives the value of the NM separation, a proxy for computational power. Note that darker grayscale pixels occur near the critical line. Adapted from Bertschinger and Natschlager (2004).

found the parameters that produced maximal computational power, as measured by the NM separation, closely coincided with those that produced critical dynamics. This suggests that computational power would be maximized near the critical point, where dynamics would be neither chaotic nor attractive, but dynamically neutral.

Dynamics

Now let us talk about dynamics in more detail, as this is the third realm where criticality may bring optimization. As we just said, there are three broad types of dynamics that any system can have: attractive, neutral and chaotic. To get a very quick intuitive picture of this, consider balls rolling down inclined surfaces, as shown in figure 4.13. In the left panel, the surface contains a valley in the middle, causing balls that are spaced widely apart at the top to flow together as they approach the bottom. Their trajectories become closer over time, so we call these dynamics attractive. In the center panel, the surface is a completely flat plane. If balls are placed apart by a given distance at the top, they will preserve this distance by the time they reach the bottom—they are not closer together or further apart. This is an example of neutral dynamics. In the right panel, the surface has a bulge in the middle, causing closely spaced balls at the top to roll apart from each other by the bottom. This is an example of chaotic dynamics, where small initial differences are amplified over time. These are qualitative descriptions; the appendix describes in more detail ways to quantify these dynamics.

Each type of dynamics has features that may be relevant for the brain. Attractive dynamics are often invoked in models of memory, where widely varying cues all converge on the same information. For example, consider these disparate items: a dollar bill, the capital city of the United States, and the first US president. In memory space, they all flow together toward the topic of George Washington. In this framework, memories are seen as states toward which the network is attracted. While attractive dynamics may be good for retrieving memories, it makes controlling the network hard because it is always being drawn toward the attractive basins. Moving the network in paths away from these basins becomes difficult.

Neutral dynamics, while typically not considered as useful for memory, allow the network to be controlled more easily. A small adjustment in a trajectory is maintained over time, without damping or amplification. Systems with neutral dynamics are often called

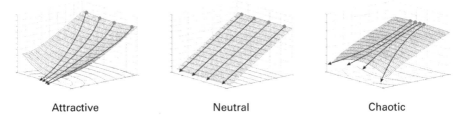

Attractive   Neutral   Chaotic

Figure 4.13
Three different types of dynamics. Balls rolling down inclined surfaces illustrate how trajectories can be brought together with attractive dynamics (*left*), remain equidistant with neutral dynamics (*middle*), or grow in distance with chaotic dynamics (*right*).

"marginally stable," meaning that they reside just at the boundary between stability and instability. To get a picture for this, consider a bicycle. It is not stable by itself and takes some time to learn how to ride. On the other hand, it is not completely unstable. If, while riding, you were to jump off it smoothly, it would continue to roll on by itself for a little while on roughly the same path. This property allows you to steer it anywhere you want it to go, with a minimum of effort. Such dynamical features could be desirable for some neural networks where steady control is more important than memory. Because neutral dynamics reside at the border of instability, they are sometimes referred to as being on the "edge of chaos." When the branching model is tuned to the critical point, it shows dynamically neutral trajectories (Haldeman and Beggs 2005), where stability and controllability can coexist.

Chaotic dynamics have been predicted by models (Van Vreeswijk and Sompolinsky 1996) and often are associated with seizures and epilepsy (Babloyantz and Destexhe 1986), as slight perturbations are typically amplified. Although chaotic dynamics are more difficult to control, there are ways of taming them through appropriately timed stimuli (Schiff et al. 1994). The variability of chaotic dynamics can be seen even as advantageous under special conditions where chaos can be harnessed to perform computations (Sinha and Ditto 1999).

Summary of Areas Yet to Be Tested
Together, these studies suggest that information storage and computational power would be simultaneously maximized at the critical point. This also would be the point where the dynamics of the cortex would be nearly balanced, preventing instability yet allowing controllability. At the time of this writing, the consequences of operating near the critical point have not yet been tested on information storage capacity, computational power, or controllability in living neural networks.

Chapter Summary

The cerebral cortex is, above all, an information processing organ. How could this, the crowning achievement of evolution, perform all its varied information processing tasks optimally? Results from the branching model predict that performance on many tasks could be optimized simultaneously near the critical point. We reviewed experiments that validated these predictions for information transmission, dynamic range, and susceptibility. Computational models show that maximal information storage and computational power should also occur near the critical point. In addition, the critical point would allow the network to be stable yet controllable. Perhaps experiments will soon test these predictions. The fact that all these information processing functions can be made to peak by tuning connection strengths shows that they depend on how neurons interact with each other. Optimal information processing is thus a collective phenomenon that emerges near the critical point.

Exercises for this chapter can be found through a link given in the appendix.

5

Universality

All science is either physics or stamp collecting.
—attributed to Ernest Rutherford

Well, it turns out there are some stamps worth collecting.
—Sydney Brenner

In very broad terms, science has two types of activities that are always going on. First, there is the collection of facts through observation and experimentation. Second, there is the organization of these facts into some sensible structure. For example, Darwin extensively catalogued features of species from his voyage on the *Beagle* and then came to the insight of organizing them through the principle of natural selection. Mendeleev studied the different properties of elements before conceiving of the law that led to the periodic table. A pattern cannot be revealed unless the facts are first known; stamps must be collected before theories can be born.

On a more mundane level, this process of jumping from facts to principles is being attempted every time a hypothesis is cast. Dale's law in neuroscience was the hypothesis that a single neuron could only secrete neurotransmitters of one type from all its synaptic terminals. Although we now know of exceptions to this law, it was a very sensible attempt to distill a general principle out of a vast array of facts about cell types and transmitter molecules (O'Donohue et al. 1985).

From this perspective, the hypothesis that the cortex operates near a critical point is another proposal to organize many disparate facts into one relatively simple principle. There are many who think that offering such a hypothesis is premature. Over a decade ago, I had a lunchtime conversation with a prominent visiting neuroscientist. He routinely publishes in the most elite journals and his work garners several thousand citations a year. When he asked me to explain the hypothesis I was working on, he listened carefully for about five minutes and then responded with "Let me be polite—I don't think any such principle could apply to the brain without more regard for all the details we first need to know." He went on to indicate that he thought it was naïve to expect such general principles in biology. I have received similar responses from other colleagues, so this reaction is not particularly unusual.

We thus have a tension between the desire to collect more information and the hope that we have collected enough already to propose an idea that could organize it. It is natural to wonder how one should proceed in the case of the criticality hypothesis for the cortex. To resolve this, we will turn to the concept of *universality*. As we mentioned before, when a system operates near the critical point, two major consequences follow. First, it has scale-free properties. In the case of the cortex, these lead to optimum information processing as we saw in the last chapter. Second, a system near the critical point should display universality; the present chapter is focused on addressing this second issue. But what is universality and how can we understand it?

We will take a short digression into physics to explain how the concept of universality first arose. It turns out that many different physical systems have phase transitions, and they behave in very similar ways when they operate near the critical point. Interestingly, this means that the important features of a system's behavior near the critical point are independent of its microscopic details. In other words, near the critical point, water, magnetic materials, avalanches in piles of sand, and cascades of neural activity in the cortex all show profound similarities. This may sound unbelievable, so we will illustrate with some concrete examples. These will build intuitions about what it means for a physical system to demonstrate universality.

After these illustrations, we will turn to consider what universality would mean in a biological system; more specifically, the cortex. If some lower-level details don't matter, then we should expect to see similar behavior in several different species. Differences in brain structure between mammals and reptiles, for example, should not hinder it. In addition, if these signatures are universal then they should appear similar at different scales, ranging from cortical microcircuits to the whole brain. Finally, if many details don't matter, then any model used to capture these features should be devoid of such details. This means that all these systems should be described by a relatively simple model. After explaining these criteria in more detail, we will then look to see if recent neuroscience data can meet these claims.

Universality in Physical Systems

Our understanding of universality has its origins in the first experiments involving phase transitions. These experiments required pressure and temperature to be controlled, so different phases of a substance could be observed easily. Baron Cagniard de la Tour in 1822 rolled a flint ball inside a "steam digester," a device much like a modern kitchen pressure cooker, filled with water. By rolling the device, he could hear the ball splashing as it passed through the liquid-vapor interface. At a very high temperature, he could no longer hear the splashing sound, indicating that the difference between the gas and liquid phases had disappeared (Berche, Henkel, and Kenna 2009). Although he had reached the critical point, it took many years before scientists appreciated the generality of this phenomenon. Similar experiments were done with other substances, each time yielding different temperatures, making it difficult to see a clear pattern.

Around 1895, Pierre Curie showed that magnetic materials lost their magnetism when they were heated above a critical temperature, later called the Curie temperature (Curie 1895). Here, the transition was not between a liquid and a gas phase, but between an ordered (magnetic) and a disordered (paramagnetic) phase. The substantial differences between Curie's experiments with magnetism and the previous experiments with liquids and gases made their commonalities seem surprising. Yet there were fundamental concepts governing both phase transitions.

To illustrate some of these common principles, let us look more closely at what happens as a piece of cold iron is heated up to beyond the Curie temperature. We will explain this using a very simple conceptual model known as the Ising model. In this model, iron can be thought of as a lattice populated with many tiny bar magnets, each with a north and south pole (figure 5.1A). These bar magnets represent quantum mechanical spins, that can point either up or down, caused by the movements of electrons. Each spin interacts with its nearest neighbors and tends to cause them to point in the same direction. Far below the Curie temperature, these nearest neighbor interactions dominate, and most of the spins point in the same direction (figure 5.1B). This ordered state leads to a net magnetization of the iron, and lets it stick to a steel refrigerator door. All the tiny bar magnets sum together to produce one powerful magnet. As the temperature is increased, thermal energy causes the spins to flip, making them point up and down at different times. The nearest neighbor interactions are still there, but they are increasingly challenged by the thermal energy that causes disorder. At a very high temperature, beyond the Curie temperature, thermal energy dominates, and the spins are equally likely to be pointing up or down (figure 5.1D). Here, it is as if all the tiny bar magnets are canceling each other; the iron no longer has any net magnetization, and it would fall off the refrigerator door. Clearly, we have an ordered phase at low temperature and a disordered phase at high temperature (figure 5.1E).

But what happens right at the Curie temperature? Here, the nearest neighbor interactions promoting order are exactly balanced by the effects of thermal energy promoting disorder; there is a mixture of local order and global disorder. Locally, there can be regions where all the spins are aligned, called magnetic domains. Globally, there are domains of many different sizes, each with an equal chance of pointing up or down (figure 5.1C).

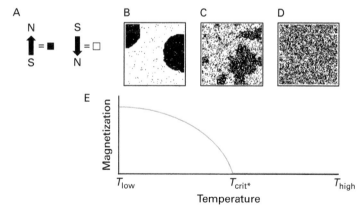

Figure 5.1
A phase transition in iron. A piece of iron is seen as a lattice with many spins. *A*, Each spin is like a tiny bar magnet with north and south poles. Pointing up will be represented by black pixels, pointing down by white. *B*, At low temperatures, nearest neighbor interactions cause spins to point in the same direction. The simulation shows large domains with a majority of spins pointing down. *C*, As temperature is increased, thermal energy competes against nearest neighbor interactions and some groups of spins form domains pointed in the opposite direction; domains of all sizes form. *D*, At very high temperature, thermal energy far exceeds nearest neighbor interactions, so local order within large and medium sized domains is disrupted. Only small domains exist. *E*, Magnetization is the sum of all the bar magnets: strong at low temperatures with ordered alignment, approaching zero as *T* moves toward the critical temperature and zero for all *T* above.

At this phase transition point, the iron behaves in ways that are remarkably like the neural network model and the gas-liquid system we discussed earlier. For example, in iron the sizes of the magnetic domains follow a power-law distribution; the smallest domains are much more common than the largest domains. This is analogous to the distribution of avalanche sizes in the neuronal model. In iron, if we flip a spin from up to down, the nearest neighbor interactions will cause an avalanche of other spins to flip. These avalanches will propagate longest and furthest when iron is at the Curie temperature and will propagate less in time and space when it is not. Now think about a sample of water: if we go in and displace a single molecule, that perturbation will displace other molecules and propagate throughout the sample. It will propagate furthest in time and space when the water is right at the phase transition point between gas and liquid. It will propagate less when the water is away from the critical point.

Although these systems are microscopically very different, they show common macroscopic behaviors. These common behaviors hint that there may be some underlying universal principles governing these systems near the phase transition point. The Ising model we have been discussing, developed around 1924 by Wilhelm Lenz and Ernst Ising (Brush 1967; Niss 2005), played a crucial role in helping scientists first to understand these universal principles. Here we will highlight two of its main contributions.

First, the Ising model illustrates the broad applicability of a phase transition and a critical point. With some adjustments, the Ising model has been adapted to describe how networks of neurons store memories (Hopfield 1982, 1984; Schneidman et al. 2006), how activity patterns in the human brain are determined by its connections (Fraiman et al. 2009), how different cell types segregate during limb development (Graner and Glazier 1992), and how people make group decisions in social settings (Galam 1997). As we will soon see, while this broad applicability may have originated with the Ising model, it is not limited to it. The Kuramoto model describes what happens in a network of oscillators as it undergoes a transition from an unsynchronized phase to a synchronized phase (Acebrón et al. 2005). The percolation model describes the changes that occur when many small, independent pathways in granular material (like coffee grounds) start to link and go from a disconnected phase to a connected phase (Domany and Kinzel 1984). Both models have been applied in neuroscience (Breskin et al. 2006; Kitzbichler et al. 2009; Benayoun et al. 2010), like the Ising model.

Second, and most pertinent to universality, the Ising model demonstrates that some lower-level details do not matter. This will take a few steps to explain. As we mentioned, the Ising model produces power laws for domain sizes and for the spread of a perturbation, among other things. Each of these power laws has a slope in a log-log plot and therefore has its own characteristic exponent, like the values α and τ we saw before. It turns out that every system that undergoes a phase transition with a critical point will have a set of exponents. These exponents are used to classify the different types of phase transitions we have discussed, like from inactive to active, from ordered to disordered, from desynchronized to synchronized, or from disconnected to connected. Most interestingly, these exponents often do not depend on many of the lower-level details of the model or system. For example, for fluids at the liquid-gas phase transition, the critical exponents are the same, independent of the chemical that makes up the fluid (Stanley 1971). In a similar manner, the critical exponents for the Ising model do not differ if we implement the model

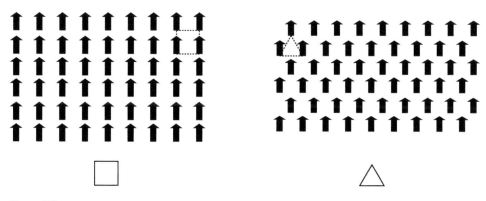

Figure 5.2
Some lower-level details don't affect the phase transition. Spins on a square lattice, *left*, or spins on a triangular lattice, *right*. In the Ising model, both systems will have the same critical exponents and the same behavior near the critical point. They therefore belong to the same universality class.

on a square lattice or a triangular one (Nishimori and Ortiz 2010) (figure 5.2). The macroscopic behavior of the system near the critical point is described fully by these exponents, and not by the lower-level details. This fact marks a radical departure from the typical reductionistic view in science that often assumes that it is the lower-level details that drive higher-level phenomena.

This way of describing systems by their critical exponents is so powerful in fact that physicists have catalogued entire groups of systems that belong to the same universality class. For example, epidemics spreading through a population, water moving through coffee grounds, and activity cascading through neural networks may all be examples of the universality class called "directed percolation." The essential behavior of this universality class can be captured with the branching model. For large system sizes near the critical point, these diverse systems will all have exactly the same critical exponents for avalanche sizes, avalanche durations, and avalanche sizes vs. durations. Another example is the "two-dimensional Ising model" universality class. Members of this class include ferromagnets, fluids like water, and the majority voter model. Their behavior can be captured by the two-dimensional version of the Ising model, and they all share the same critical exponents for susceptibility and other quantities like the specific heat.[1]

The notion of universality is one of the greatest triumphs of modern physics (Wilson 1979). It shows that large-scale, emergent behaviors are not necessarily driven from the bottom up, by lower levels. It also shows that some complex systems can be understood through relatively simple models or equations. Universality moves us away from the frightening morass of endless facts and puts us a little closer to the day when we could discover a few equations to describe the brain.

Universality in the Cortex: Indicators

Given this background in universality, let us now switch to consider the cortex again. We will explore what it would mean for this type of understanding to be achieved with the brain—to find some general principle about its operation that would apply to reptile brains as well as mammalian brains; to three-layered as well as six-layered cortex.

For this to work, we would of course need to have clear indicators of when a network is operating near the critical point. We have already discussed these, and they include evidence of distributions that approximately follow power laws and evidence that the exponents from these distributions satisfy an exponent relation. These indicators would allow us to clearly delineate networks operating near the critical point from those that are not, in different species and at different scales. With this in hand, we would be ready to test if the generality of this proposed principle is true. But we would need to see three additional things to demonstrate a universal principle:

Indicators seen across species. These indicators of being near the critical point should be seen in many different species. They should not only appear in recordings from rats, but also primates, humans and possibly other less related animals like fish and reptiles. This would support the idea that the principle is generic.

Indicators seen across scales. If brains near the critical point show universal operating principles, these principles should be detectable at many different scales. This means that data collected from micro-scale recordings from hundreds of spiking neurons, as well as coarser-grained signals from local field potentials (LFPs), optical recordings across large patches of cortex, and even whole-brain data from magnetoencephalography (MEG) should all produce approximate power laws with exponents that satisfy an exponent relation. Seeing a consistent picture across all these scales and across different techniques would argue in favor of some universal principle.

Described by a simple model. As we mentioned, one relatively simple model should be capable of describing all these data sets from different species and from different scales. This notion was concisely framed by Jim Sethna, Karin Dahmen, and Chris Myers in a review article:

Because these systems exhibit regular behavior over a huge range of sizes, their behavior is likely to be independent of microscopic and macroscopic details, and progress can be made by the use of simple models. The fact that these models and real systems can share the same behavior on many scales is called universality. (Sethna, Dahmen, and Myers 2001)

Sethna and colleagues explain how a broad array of systems that operate near the critical point share common dynamics. From the crumpling of a piece of paper, to avalanches in a pile of sand, to earthquakes, all obey a similar set of equations. Given this, it makes sense that models used to describe these systems will not depend crucially on details found only at a particular scale—the details of paper crumpling differ from the details of slipping tectonic plates. Instead, relatively simple models will be sufficient. To translate this to the realm of neuroscience, if something like the critical branching model can describe neuronal avalanches in multiple species and across multiple scales, then we will have evidence supporting universality. Let us now see if these three criteria can be met by data from recent experiments.

Indicators Seen across Species

As we discussed earlier, one of the best indicators of operating near the critical point is that the exponents from the data satisfy an exponent relation. Experiments by Fontenele and colleagues examined this with data from awake, freely moving mice (Fontenele et al.

2019). Interestingly, they found that the exponents could differ from mouse to mouse, and even from time to time within the same mouse. On the surface, this sounds like it would be bad news for any theory claiming universality; indeed, we will need to return to this later. However, when they plotted the exponents for avalanche duration (α) and size (τ) from each experiment, they found that they fell along a line with slope of 1.28 (diamond icons in figure 5.3). This indicated that the data not only satisfied the exponent relation, but that they all shared the same exponent, γ. Recall that γ describes how average avalanche size scales with duration ($S \sim T^\gamma$). Equally interesting, though, is that when they plotted data collected from other labs using anesthetized rats, anesthetized monkey, or even turtle brains or cultured slices of cortex, they found that these points fell close to the same line as well. This relationship has since been found by other groups also (Mariani et al. 2021).

This result is remarkable for several reasons. First, it is rare to see biological data produce such a clear pattern. The points are not widely scattered around a line, faintly suggesting a trend among them; most of them lie right on, or very close to, the line. This gives the strong impression of an underlying order. Second, these data points come from widely different species and include both awake and anesthetized animals. It does not seem to matter if the animal has a three-layered cortex (the turtle) or a six-layered cortex (the mammals), or whether it is awake or asleep. This underlying order transcends those details. Third, some of these data points are taken from the same animal at different times. At one time the animal might be quietly sitting, while at another it might be walking around. While the neuronal avalanche distributions could change exponents under these conditions, the relationship between these exponents stayed the same. Under varying behaviors and stimulus conditions, the networks move along the dashed line, where they remain close to the critical point. This plot thus suggests there are mechanisms to maintain operation close to the critical point even as input to these networks is changing. We will elaborate on this later when we address homeostasis in chapter 6.

Extending the range of species even further, work by Ponce-Alvarez et al. (2018) found that zebrafish larvae also show signatures of operating near the critical point (figure 5.4). Zebrafish are a widely used model organism in part because they are nearly transparent as

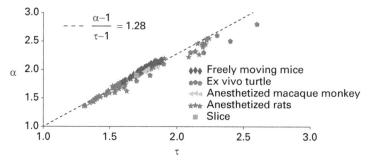

Figure 5.3
Brains of turtles, mice, and monkeys all operate near the critical point. Each token (diamond, pentagon, triangle, star, square) represents a single experiment where the avalanche size exponent, τ, and the avalanche duration exponent, α, are plotted. Notice that all tokens lie on or close to the dashed line given by the exponent relation $(\alpha - 1)/(\tau - 1) = \gamma$, where γ is 1.28. In each case, the network producing the data is close to the critical point. Modified from Fontenele et al. (2019).

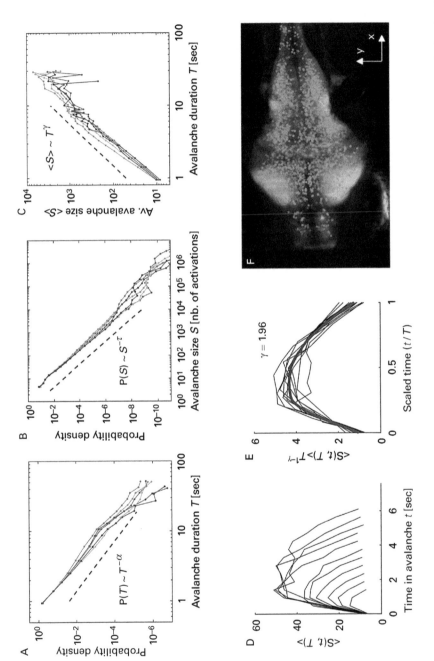

Figure 5.4

Zebrafish larvae brains operate near the critical point. Calcium imaging at nearly single neuron resolution of the entire zebrafish larva brain. *A*, Avalanche duration (*T*) distributions from six data sets approximately follow a power law with exponent α. *B*, Avalanche size (*S*) distributions approximately follow a power law with exponent τ. *C*, Average avalanche size, ⟨*S*⟩, plotted against duration, *T*, also follows a power law, with exponent γ. *D*, Average avalanches of 11 different durations. *E*, These avalanches approximately collapse onto the same scaling function and satisfy the exponent relation. *F*, Image of zebrafish larva brain, taken through its transparent skull. Gray dots indicate avalanche initiation sites. Modified from Ponce-Alvarez et al. (2018).

larvae. With genetically encoded calcium indicators and rapid light sheet microscopy, investigators can noninvasively image activity in nearly every neuron in their brains. Moreover, this imaging can be done while zebrafish are awake and giving motor responses to visual stimuli. Ponce-Alvarez and colleagues found that spontaneous activity consisted of neuronal avalanches that followed the exponent relation, but that sensory inputs and tail movements would transiently cause dynamics to depart from near the critical point (Ponce-Alvarez et al. 2018). After these departures, activity returned to near criticality within about 10 seconds. Among the many interesting findings of this study, it suggests that operating near the critical point may have appeared very early in evolution, at least as far back as fish.

At the scale of hundreds of individual spiking neurons, then, the data present a consistent picture that brains of many species operate near the critical point. Let us now see if this evidence extends to larger scales as well, where a suite of different methods has been employed to measure the summed or averaged activity of increasingly larger groups of neurons.

Indicators Seen across Scales

Fortunately, the topic of criticality in brains has been around long enough that it has been investigated with several techniques at different scales. Figure 5.5 shows three representatives of this work spanning scales from 1 mm up to 150 mm. In each case, avalanche size distributions follow a power law with about the same negative slope of 1.5. To further highlight the independence of scale, in each example the investigators recorded data from at least three different configurations of their recording system. More specifically, with the LFP and MEG signals, data were taken from subsets of the full array. For the voltage sensitive fluorescent protein (VSFP) signals, the image of the cortical hemisphere was sampled entirely, but with three different pixel cell sizes. Regardless of these different sampling methods, the results were approximately the same. This means that the dynamics in a tiny patch of cortex are statistically similar to those in the entire cortical mantle (Agrawal et al. 2019).

Although it is important to note that these represent networks of very different sizes, it is interesting to realize that each element of these networks contains very different numbers of neurons as well. Just making rough estimates, each negative LFP spike is the average from tens to hundreds of neurons in the vicinity of the electrode; each pixel in the VSFP image gives the activity of several thousand neurons, and each MEG sensor can capture the activity of perhaps many millions of neurons.

Besides these power laws, other indicators also suggest these networks were operating near the critical point. For the LFP data, the branching ratio was approximately 1 (Beggs and Plenz 2003). For the VSFP data, a measure called kappa, indicating the extent to which the avalanche size distribution deviates from a perfect power law, indicated that the data were nearly ideal (Scott et al. 2014). For the MEG data, the branching ratio was also found to be very close to the critical value of 1 (Shriki et al. 2013). Taken together with the previous results from spiking networks, these show nearly critical dynamics govern small networks consisting of a few hundred cortical neurons, all the way up to the entire human cortex, with its estimated 16 billion neurons (Herculano-Houzel et al. 2015)—a span of over seven orders of magnitude.

Because these results do not depend on one method alone, they are less vulnerable to the criticism that they could be an artifact of a specific recording technology. For example, it

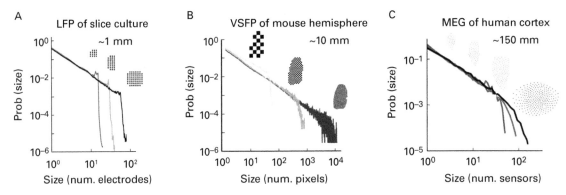

Figure 5.5
Similar avalanches across scales, across methods. *A*, Distributions of avalanche sizes taken from local field potential (LFP) signals that crossed a negative threshold, in cultured cortical slices from rats (Beggs and Plenz 2003). Width of slice is on the scale of ~1 mm. To show invariance to sampling, data were taken from one quarter array (left icon), half array (middle) and full array (right). Distributions from all three have the same slope of −1.5. *B*, Distributions of avalanche sizes from optically recorded voltage sensitive fluorescent proteins (VSFP) in one hemisphere of mouse cortex (Scott et al. 2014). Width of mouse brain is ~10 mm. Data were coarsely sampled (left icon, each pixel 1 mm²), normally sampled (middle, pixels 270 μm²), or densely sampled (right, pixels 70 μm²). Distributions from all three scales have nearly the same slope of −1.5. *C*, Avalanche size distributions from magnetoencephalography (MEG) recordings from human cortex (Shriki et al. 2013). Width of human brain is ~150 mm. Data were sampled with subsets of the sensor array of various sizes (icons from left to right), and all produced distributions with approximately the same slope of −1.5. Overall, similar results are seen across scales ranging from 1 mm to 150 mm, and from very different methods.

has been argued that the power-law frequency spectrum observed in LFP signals could be caused by the filtering properties of the extracellular space on the local electric field (Bedard, Kroeger, and Destexhe 2006). This argument would not necessarily apply to individual spike data or recordings from calcium imaging. The diversity of methods, all pointing toward the same conclusion, suggests that nearly critical behavior across scales is a robust finding.

Described by a Simple Model

Now we can turn to the issue of whether a simple model can account for the nearly critical dynamics observed across scales. Let us review why we expect this will be possible. Activity cascades across the whole brain are shaped by many macroscopic features, like the distance between cortical areas and the density of fiber bundles between them. Analogously, neuronal avalanches in cortical circuits will be determined by microscopic features like the layered structure of cortex and the number of synapses between layers and cell types. Yet if activity at both scales shows signatures of being near the critical point, then it suggests that they share some common principles of operation, despite their differences in biological details. This is the reasoning behind universality—that some model that is independent of information about fiber bundles or cortical layers will be sufficient to capture the essential dynamics of cortical activity across all seven orders of magnitude in scale. Because such a model will be independent of details, it should be relatively simple and convey a central concept.

We have already seen that the branching model can describe the power laws, as well as the exponent relation and avalanche shape collapse, for data from networks of spiking neurons. Can it do the same for data at other scales as well? There is some reason to suspect that it might not be able to. Astute readers may have noticed that not all the data that satisfy the exponent relation share the same exponent γ. For networks of spiking neurons, we have γ ≈ 1.3, while for zebrafish larvae we nearly have γ ≈ 2.0. Although it might seem that this difference in the exponent γ negates our claim of universality, there is a simple way we could explain a wide range of data with the branching model. To understand this, we will first need to take a brief detour to describe what the exponent γ can tell us about a neural network.

From our previous discussions in chapter 3, we know the exponent γ relates the avalanche duration (for a given size) to the avalanche size through this equation: $S = T^{\gamma}$. Let's consider three different values for γ. First, what if γ were 1? In this case, the avalanche size would merely be given by its length ($S = T^{1}$). An avalanche of duration 5 would only be of size 5, meaning that at every time step, only one neuron would be active. Such avalanches would always have a strictly chainlike structure, and their shapes would be flat for all lengths (figure 5.6B), not showing the characteristic inverted parabola we saw before in figures 3.6 and 3.13. Since the exponent is 1, these avalanches would be one-dimensional, proceeding along a line. Second, what if γ were 2? Here, the avalanches would fan out much more profusely, growing in size as the square of their duration (figure 5.6D). An avalanche of length 2 would have a size of 4, and an avalanche of length 3 would have a size of 9, and so on. They would be two-dimensional, with shapes that could not be flat. Third, consider γ = 1.3, an exponent between the whole numbers 1 and 2, and what we have seen in data. Here again the avalanche shapes cannot be merely flat as in the one-dimensional case, but they cannot arc up as highly as in the two-dimensional case (figure 5.6C). Now they have a fractional dimension that is not a whole number, and for this reason they can be called "fractal." As we mentioned before, fractals have the property of being self-similar across scales. From this discussion, it should be clear that the exponent γ is highly informative about the structure of avalanches. More specifically, it can tell us about the pattern of functional connectivity between neurons on which the avalanches spread.

Since the exponent γ is related to how the avalanches fan out within the network, it should come as no surprise that changing the density of connectivity within a network model can change the value of γ that is measured in the output of simulations. Theory predicts that a network with all-to-all connectivity, where each unit (e.g., neuron, cortical area) is connected to every other unit, will have an exponent of γ = 2. This is nearly the case in the data from the zebrafish larvae, where γ ≈ 1.85 (Ponce-Alvarez et al. 2018) and may be the case for LFP data collected in monkeys, where γ ≈ 2.0, once the impact of background gamma oscillations has been taken into account (Miller, Yu, and Plenz 2019). Sparser connectivity will lead to lower values of γ, like what we see in data from spiking networks (Fontenele et al. 2019). The lowest possible limit for a critical network would be γ = 1.0, where connectivity would be chainlike; below this, even chainlike avalanches would be impossible. At the other end, theory predicts that values of γ above 2 would be incompatible with critical avalanches because they would expand too quickly.

By merely having different connectivity densities, then, the simple branching model can match the experimental data from diverse scales. When the branching ratio is tuned to the critical value of 1 and the model is run on these different networks, they will all produce avalanches whose exponents satisfy the exponent relation. For example, a model with all-to-all connectivity would give us exponents γ = 2, α = 2, τ = 1.5, which approximately agree with

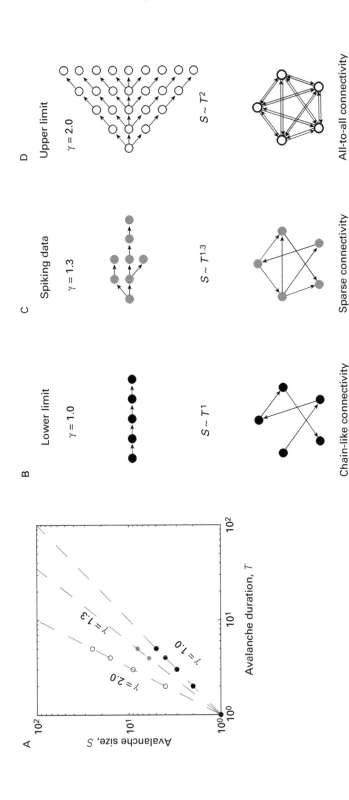

Figure 5.6

The branching model can capture data with different values of the exponent γ. *A*, Average avalanche size for a given duration, S, plotted against avalanche duration, T, for the case where $\gamma = 1$ (black), $\gamma = 1.3$ (gray), and $\gamma = 2$ (white). *B*, Avalanches for $\gamma = 1$ must all be chainlike. In this case, they are one-dimensional. *C*, Avalanches for $\gamma = 1.3$ have an intermediate fan out that is fractional, giving the avalanches a fractal dimension. *D*, Avalanches for $\gamma = 2$ must fan out more rapidly and are two-dimensional. Note that the branching model simulated in figures 3.4 and 3.6 has $\gamma \approx 1.3$.

data from LFP recordings in monkeys (Miller, Yu, and Plenz 2019). These are the exponents for what is called the "mean-field directed percolation" universality class. Mean field just signifies that all the neurons are connected to each other, and the universality class means that there are several systems that can be described by the same set of exponents, as we mentioned earlier. A model with sparser connectivity can give us different exponents like $\gamma = 1.33$, $\alpha = 1.61$, $\tau = 1.46$, which approximately agree with spike data from cortical cultures (Fosque et al. 2021). The branching model with adjustable connectivity then would be an example of a simple model consistent with the requirements of universality.

Just as the branching model seems to answer some questions, it also raises others. For example, if all the data points in figure 5.3 seem to have the same exponent γ, then why do they have different values of α and τ? At first glance, this seems to suggest that they are not in the same universality class. To understand how this might still be the case, we will need to consider how external inputs to a network may change the exponents that are effectively measured. This will be covered later in chapter 7, which is dedicated to the topic of quasicriticality.

Before finishing, it seems important to reflect on the implications of this result. Back in chapter 2, when we studied stadium wave, we concluded that its speed, size, and reflective properties did not depend on many lower-level details like the precise distance between people or what language they spoke. In a similar manner, neuronal avalanches that occur near the critical point have properties that do not depend on whether a brain is reptilian or mammalian, or whether the cortex has three layers or six. Being near the critical point brings with it the emergent phenomena of enhanced susceptibility, increased dynamic range, optimal information transmission, and possibly optimal information storage and computational power. These properties are universally beneficial to all brains, regardless of species.

Chapter Summary

Most broadly, there are two main consequences for a neural network that operates near the critical point. First, it will have scale-free properties; in chapter 4 we saw that those can be linked to optimal information processing. Second, it will exhibit universality; this is the topic we explored here in chapter 5. To build intuitions about universality, we discussed the Ising model; despite its simplicity, it can accurately describe the critical behavior of a piece of iron, diverse fluids, and a model of collective human interactions. We then turned to the cortex and noted that for a theory to be universal there, it should accurately describe the data across species and across scales. We reviewed data supporting both these points. The last requirement for universality is that one simple model be able to describe the phenomena observed across species and scales. We showed that the branching model can do this when connectivity density is adjusted appropriately. The idea that emergent phenomena have properties that are independent of some lower-level details is therefore applicable to the cortex. Universality is not something that the cortex uses for its own operation. Rather, it is something that helps us understand the cortex. It allows us to use simple general principles to describe its operation without having to know millions of facts about it.

Exercises for this chapter can be found through a link given in the appendix.

III

Future Directions

6

Homeostasis and Health

The coordinated physiological processes which maintain most of the steady states in the organism are so complex and so peculiar to living beings—involving, as they may, the brain and nerves, the heart, lungs, kidneys and spleen, all working cooperatively—that I have suggested a special designation for these states, homeostasis. The word does not imply something set and immobile, a stagnation. It means a condition—a condition which may vary, but which is relatively constant.
—Walter Bradford Cannon, who is credited with coining the term "homeostasis"

A brain can improve till it fits its environment.
—W. Ross Ashby, author of *Design for a Brain*

There is an experiment in our advanced undergraduate physics lab called "critical opalescence." The idea behind it is to bring water to the critical point, where the difference between liquid water and gaseous steam disappears. To get there, the water must be pressurized to about 220 atmospheres and heated to 374.2°C (705.6°F). For safety, this takes place within a small chamber that is tightly sealed, with only a tiny observation window made of thick glass. Long before reaching the critical point, the liquid water is clearly visible at the bottom of the chamber, with the steam above it; both are transparent. As pressure and temperature are increased to approach the critical point, the boundary between water and steam becomes unclear, and the chamber looks cloudy and grayish. This loss of transparency is caused by correlations between water molecules whose length becomes large enough to match the wavelength of light and scatter it. Right at the critical point, the correlations grow as big as the chamber itself, and the sample becomes milky white because all wavelengths are scattered. But it only looks this way for a fraction of a second. That was it—the critical point—and if you blinked you probably missed it. In vain, students tap the glass and lower the pressure by just enough to try get back there, but they usually fail. They successfully passed through the critical point, but they didn't land on it, because that is virtually impossible.

We probably should not think that reaching the critical point in the cortex would be any easier. In fact, it should be much harder. Sensory inputs to the cortex are constantly changing, synaptic plasticity causes connections to vary in strength, and massive pruning

of synapses occurs during development. On top of this, there are sometimes large perturbations caused by strokes, concussions, and fevers. For the cortex to be nearly critical
under all these conditions, it would have to be tuned there by some mechanisms to counteract the forces pushing it away from criticality. Even then it seems that hitting the critical point exactly and staying there would be unlikely. But as we have seen, just being near
the critical point is often enough, in a finite cortex, to confer on it nearly optimal information processing. This then raises the question of how the cortex could get close to the critical point and stay near there.

One way to approach the critical point, even in the presence of perturbations, would be
through negative feedback. A thermostat works on this idea—when the house is too hot,
it turns on the air conditioning; when the house is too cold, it turns on the heater. As long
as the house is within some acceptable zone, say $22 \pm 2°C$, then no action is taken. Perhaps the cortex could homeostatically regulate itself to hover near the critical point in a
similar manner. Do we have any evidence of this? And if we do, does it reveal the biological mechanisms that could accomplish such critical homeostasis?

In this chapter, we will review experiments on homeostatic regulation toward the critical point that were done with rats, turtles, humans, and cultures of neurons. We will see
that while they provide a consistent picture that some type of critical homeostasis exists,
they do not definitively reveal the mechanisms that might be driving homeostasis. This is
still a relatively new area of research, with exciting open questions. Fortunately, there are
several computational models to suggest how homeostasis is accomplished, so there is no
shortage of testable hypotheses.

If there is homeostasis around the critical point, and if the critical point is nearly optimal for information processing, then it would make sense that departures from the critical
point should be associated with impaired cognition or poor neurological health. In this
chapter we will also examine studies that suggest this. Again, this is a relatively new but
rapidly growing area of research.

More broadly, this chapter, as well as chapter 7 on quasicriticality and chapter 8 on the
cortex, are devoted to the frontier areas of research surrounding the critical point. For
these topics we have enough information to report on how the hypothesis of criticality is
influencing current research and we can begin to see the outlines of future research, even
if we do not have the final answers yet.

Homeostasis toward the Critical Point after a Major Perturbation

The idea of homeostasis in the nervous system is not new and was hypothesized at least as
far back as 1926 by Walter Bradford Cannon, the person who is credited with coining the
term "homeostasis" (Fleming 1984). Revealing the mechanisms of homeostasis in the brain
occurred much later, in seminal work by Gina Turrigiano, Sacha Nelson, and colleagues
(Turrigiano and Nelson 2004). They showed that the synaptic strengths of neurons, as
well as their firing rates, are homeostatically regulated (Turrigiano et al. 1998). In experiments, they exposed cultured neurons to the toxin from puffer fish, tetrodotoxin, which
blocks sodium channels and hence action potential firing. After about a day of exposure,
the neurons responded by producing more synaptic glutamate receptors to compensate,
causing their firing rates to rise back to baseline levels. They called this phenomenon

"synaptic scaling" because it had the effect of multiplying all synaptic currents by the same constant amount. Similarly, when networks of cultured neurons were bathed in bicuculline, which blocks inhibitory neuron transmission, excitatory neurons began to fire at much higher rates. After about a day, the excitatory neurons responded—this time by reducing the number of synaptic glutamate receptors, thus reducing their firing rates back toward baseline levels. Although these mechanisms are at the synaptic level, their discovery made it plausible to expect that homeostasis at a network level could also exist. Computational models incorporating firing rate homeostasis and synaptic scaling predicted that living neural networks should return to the critical point after perturbations (Effenberger, Jost, and Levina 2015), but experiments testing these predictions had not been done.

To test investigate these ideas, Zhengyu Ma, Keith Hengen, and colleagues (Ma et al. 2019) sought to observe how visual cortex in freely behaving rats would respond to a strong, long-lasting perturbation. They implanted microelectrode arrays to record activity in both left and right visual cortex. These arrays allowed them to continuously monitor spiking activity for over 200 hours, giving them plenty of time to track how any homeostatic changes might gradually unfold. When one of the eyes was sutured shut (figure 6.1D), depriving the contralateral cortex of input, they noticed a decline in the firing rates after about 24 hours. No such decline occurred in the control cortex, where the eye was left open. As expected, the firing rates in the manipulated cortex began to return to presuture levels after about 60 hours (figure 6.1E). This recovery suggested that the cellular and synaptic mechanisms of homeostasis that were previously observed only in cultured dishes were also active in the intact brains of behaving animals (Hengen et al. 2013). This important work established the generality of these homeostatic mechanisms.

Later, they went back and reanalyzed the data using the framework of criticality (Ma et al. 2019). To do so, they turned to the exponent relation and introduced a new measure they called the distance to criticality coefficient, DCC. Here, the DCC is just the magnitude of the difference between the left half of the exponent relation and the right half:

$$\left| \frac{\alpha_{\text{Fitted}} - 1}{\tau_{\text{Fitted}} - 1} - \gamma_{\text{Fitted}} \right| = \left| \gamma_{\text{Predicted}} - \gamma_{\text{Fitted}} \right| = DCC,$$

where α_{Fitted}, τ_{Fitted}, and γ_{Fitted} are the exponents produced by fitting straight lines to the power-law plots (figures 6.1A, B, C), and $\gamma_{\text{Predicted}}$ is the value of γ that is predicted by the fraction in the exponent relation, $\frac{\alpha_{\text{Fitted}} - 1}{\tau_{\text{Fitted}} - 1}$. Recall that if a system is exactly at the critical point, the exponent relation should be satisfied, so the DCC should equal zero. Thus, the extent to which this relation is not satisfied can be used to measure how far the system is away from the critical point. Ma and colleagues (Ma et al. 2019) recorded neuronal avalanches for 24 hours and extracted the exponents α_{Fitted}, τ_{Fitted}, and γ_{Fitted}; they found the DCC to be near 0.2 in the rats before the manipulation began (figure 6.1F). In addition to the DCC, they also tracked the branching ratio, σ, using the new accurate measures developed to counteract bias from subsampling (Wilting and Priesemann 2018). Before suture, σ was very close to the critical value of 1. Together, these suggested that the seven rats used in their study were all operating close to the critical point before the manipulation.

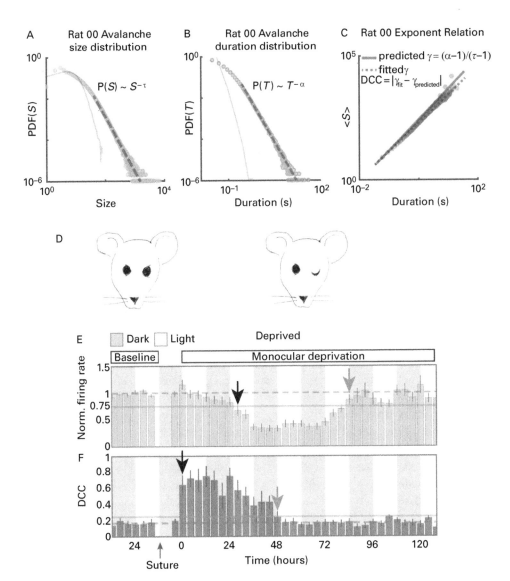

Figure 6.1

Homeostatic return toward the critical point. *A*, Avalanche size distribution from a representative rat. Dashed line shows power-law fit with exponent τ; gray curved line shows distribution from shuffled data. *B*, Avalanche duration distribution from the same rat, with exponent α. *C*, Average avalanche size for a given duration, plotted against duration. Here, the fitted exponent is γ_{fit}, dashed line, which is compared to the exponent predicted by the exponent relation, $\gamma_{\text{predicted}}$, solid line. The magnitude of the difference (absolute value) between these exponents was called the deviation from criticality coefficient, DCC. *D*, Rats recorded in baseline condition had both eyes open. During monocular deprivation, one eye was sutured shut, depriving the contralateral primary visual cortex of input. *E*, About a day and a half after deprivation, normalized firing rates ($n=7$ rats) in primary visual cortex significantly declined (black arrow). They recovered a little more than two days later (gray arrow). Gray and white background shading alternates every 12 hours to indicate dark and light periods, respectively. *F*, Deviation from criticality, DCC, grew one night after deprivation (black arrow) and recovered to baseline levels after two days (gray arrow). Control groups had no significant changes to firing rates or DCC (not shown). Modified from Ma et al. (2019) and Beggs (2019).

Next, they sutured one eye shut in each of the rats; about 12 hours after suture, the DCC significantly increased and σ significantly decreased. This indicated that the affected visual cortex was no longer close to being critical; the control cortex showed no significant changes. Even though the eye remained sutured, the DCC and σ recovered to presuture levels about 48 hours later (figure 6.1F), clearly showing evidence for homeostasis back toward the critical point. This result was notable because it demonstrated that this homeostasis existed in vivo, even in the face of a severe, long-lasting perturbation.

But the timing of this critical homeostasis was also unexpected and revealing. The DCC and σ both changed 12 hours before the firing rates dropped, and they recovered 36 hours before the firing rates recovered (figure 6.1E, F). This suggested that critical homeostasis was not merely identical with firing rate homeostasis. In fact, the distance from criticality was an earlier indicator of the perturbation and recovery than the firing rate was. This finding raised the possibility that critical homeostasis was more primary, and that firing rate homeostasis was following it.

To understand what might be happening, Ma and colleagues (Ma et al. 2019) constructed a computational network model. It was populated with excitatory and inhibitory neurons and contained three known mechanisms for self-organization in cortical circuits: firing-rate homeostasis, synaptic scaling, and a version of plasticity called STDP[1] that strengthened synapses when activity in two connected neurons was nearly coincident (Bi and Poo 1998; Abbott and Nelson 2000). After a wide parameter search that included over 400 models, they found that only a small fraction of parameter combinations could reproduce the data. Intriguingly, these successful models all pointed to the connectivity and plasticity of inhibitory neurons as the most plausible candidate for maintaining proximity to the critical point. More specifically, their models suggested that synaptic plasticity in inhibitory neurons brought the network back toward the critical point, while homeostatic plasticity in excitatory neurons stabilized firing rates.

These predictions have yet to be experimentally tested, so the mechanisms underlying this network homeostasis toward the critical point remain an open question. Another issue that this work did not address is what would happen if a positive perturbation, that would make the network supercritical, had been applied. Would homeostasis then bring it back down toward the critical point? To see the answer to this question, let us now turn to another study that explored the relationship between sleep and the critical point.

Sleep and Homeostasis toward the Critical Point

We all know how awful we feel when we don't get enough sleep and how rejuvenating it is to get even one night of uninterrupted, deep sleep. Christian Meisel and colleagues sought to explore if lack of sleep would cause the human brain to move away from the critical point, and if restorative sleep would bring it back (Meisel et al. 2013). In their study, they recruited human subjects who were willing to stay awake continuously for up to 40 hours. Since it has been reported that sleep deprivation can make a person more susceptible to seizures (Malow 2004), they hypothesized that sleep deprivation would lead their subjects to edge toward being supercritical. They were also interested in seeing if a good night's sleep would counteract this.

To collect data, they applied electroencephalography (EEG) arrays to the scalps of their subjects (figure 6.2A). These record brain activity at very high temporal resolution (millisecond range), but at relatively low spatial resolution, as the electrical signals must pass through the skull, making it difficult to localize their origins. Nevertheless, EEG signals are commonly used to diagnose neurological conditions like epilepsy (Smith 2005) and can measure responses to cognitive stimuli (Kanda et al. 2009). To extract neuronal avalanches, Meisel and colleagues (Meisel et al. 2013) applied a threshold of −4 standard deviations of the signal; waveforms that crossed this threshold were marked as events. They considered an avalanche to be a consecutive sequence of suprathreshold events, bracketed by no events at the beginning and the end.

When they plotted the avalanche size distribution for subjects before sleep deprivation, they found that it approximately followed a power law with a cutoff near 28, the number of channels they had in their array (black line in figure 6.2B). After sleep deprivation, however, this distribution had a noticeable bump near 28 (arrow on gray line in figure 6.2B), indicating an increase in avalanches whose size equaled or exceeded the size of the array. Such a

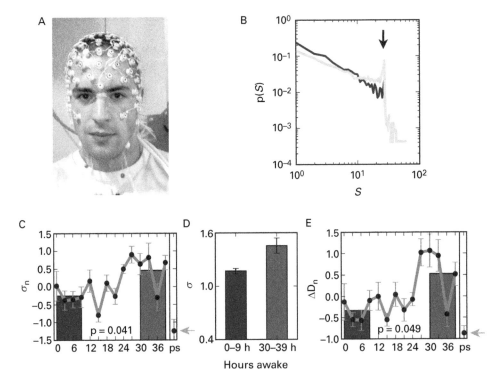

Figure 6.2
Sleep regulates proximity to the critical point. *A*, Example of an EEG array placed on the scalp of a human subject. *B*, Avalanche size distribution for control subjects, black line, and sleep-deprived subjects, gray line. Note sharp peak in distribution of sleep-deprived subjects (black arrow), indicating move toward supercritical regime. Both distributions drop near 28, the number of channels in the EEG array. *C*, Deviation of branching ratio from critical value of 1 ($\sigma_n = \sigma - 1$) measured every 3 hours significantly increased as subjects stayed awake longer but dropped after recovery sleep (ps, gray arrow). *D*, Average branching ratio (σ) for hours 0–9 and 30–39 of the experiment; it significantly increased with sleep deprivation. *E*, Deviation of avalanche size distribution from an ideal power law (ΔD_n) also increased with sleep deprivation, indicating shift toward supercritical regime, and dropped after recovery sleep (ps, gray arrow). Adapted from Meisel et al. (2013).

bump is reminiscent of what occurred in the avalanche size distributions of cortical slices and slice cultures when they were treated with picrotoxin (Beggs and Plenz 2003; Shew et al. 2009), which blocks inhibitory neurotransmission (figure 4.7A). It also occurs when the branching model is tuned to the supercritical regime (Haldeman and Beggs 2005; Pajevic and Plenz 2009). Consistent with this, when Meisel and colleagues (Meisel et al. 2013) estimated the branching ratio in their subjects after sleep deprivation, they found that it significantly increased over time (figures 6.2C, D). Taken together, these data show that sleep deprivation can cause human brain networks to deviate in the direction of supercriticality.

The next question was to find out if restorative sleep could bring them back. Interestingly, it did, but not just to the critical point. After sleep, the branching ratio dropped significantly, and even reached levels well below what it had been before the sleep deprivation began (figure 6.2C, gray arrow). The same was true of the avalanche size distributions—after restorative sleep, they not only had no hump, but they were curved and did not closely follow an ideal power law. This can be seen in figure 6.2E (gray arrow), where the deviation from an ideal power law, measured as ΔD_n, dropped below levels before sleep deprivation. Very low branching ratios and downwardly curving avalanche size distributions (in log-log plots) are indicative of subcritical dynamics (recall figure 3.10A). Certainly, there was movement back toward the critical point, but perhaps there is some overshoot if the sleep deprivation is large. What is not known is how the data would look if subjects were given more mild sleep deprivation. Again, this area of research remains active and open, with many new questions to be answered.

So far, the two perturbations we have seen have been rather extreme: eye suture and extended sleep deprivation. While critical homeostasis is active in both cases, we have not yet discussed evidence that it is also active under less extreme conditions. For that, let's turn to the next experiment.

Sensory Adaptation toward the Critical Point

Daily life is filled with events that seem to throw us out of equilibrium, if only transiently. A loud noise comes from the kitchen when a glass drops and shatters; we are bathed in a flood of bright light when we exit a dim movie theater into a sunny day; our children blindfold us to taste some chocolate milk and give us orange juice instead. All these experiences could be considered milder forms of perturbations. Is there any evidence that brains are pushed away from the critical point by events like these?

As we know from previous chapters, Woodrow Shew is an experienced researcher in the field of brain criticality. He teamed up with Ralf Wessel, who has a long track record of working with turtle brains, to conduct an experiment that tells us much about how the cortex responds to mild, transient perturbations (Shew et al. 2015). For their work, they used a turtle eye cup preparation (figure 6.3A) that Wessel and colleagues developed (Saha et al. 2011). With this they were able to access the neurons in the cortex by unfolding it and lowering a 96-channel microelectrode array (MEA). These extracellular electrodes picked up LFP signals in response to movies of natural scenes that they focused onto the eye cup.

Before a movie started, there was relatively little ongoing activity. This quickly changed at the onset of the movie, when many large neuronal avalanches began (figure 6.3D). After

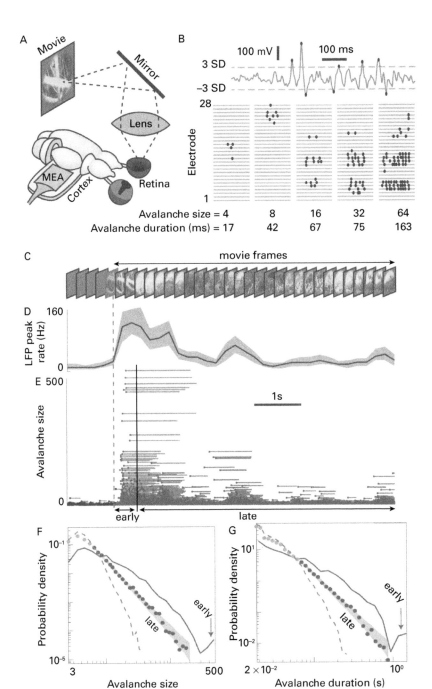

Figure 6.3

Adaptation toward the critical point after visual stimulation. *A*, Turtle eye cup preparation, showing how the movie was focused on the retina and how cortex was unfolded so that 96 channel multielectrode array (MEA) could record from contralateral cortex. *B*, Identification of avalanches. Only local field potential (LFP) activity that crossed a threshold of ±3 standard deviations was considered. Several example avalanches are shown. *C*, A movie of natural scenes was presented to the retina after a period of no stimulation. *D*, Recordings from visual cortex showed a high rate of negative LFPs at movie onset (vertical dashed line) that declined after about 1 second. *E*, Sizes of recorded avalanches plotted against time from start show that the largest avalanches occurred early, within about 250 ms of movie onset. Vertical black line divides the early from the late period. The start of each avalanche is indicated by a dot, followed by a line for the duration of the avalanche. Note that the largest avalanches start early. *F*, The avalanche size distribution from the late period approximately followed a power law (filled circles). In contrast, avalanches from the early period had larger sizes (solid gray line) and the tail of their distribution showed a positive hump (arrow) suggesting supercritical activity. Dashed line shows the distribution expected by chance (shuffled data). *G*, Avalanche duration distribution also approximately followed a power law in the late, but not in the early period. Overall, the turtle visual system adapted from a supercritical phase toward the critical point as visual stimulation continued. Modified from Shew et al. (2015).

about 500 milliseconds, avalanches continued in response to the movie, but they decreased in size. When they plotted the size and duration distributions for early avalanches initiated in the first 500 milliseconds, they found that they did not follow a power law closely (figure 6.3F, G). Rather, they showed an upward bump at the tail, like what we saw previously in the sleep deprivation study, suggesting that activity was slightly supercritical. In contrast, late avalanches that began after the first 500 milliseconds produced distributions that satisfied the statistical requirements of power laws. The exponents from these power laws also adhered to the exponent relation, showing that they were produced very close to the critical point.

This work again supported the idea that cortical networks homeostatically adjust toward the critical point, but here the perturbation caused by the onset of stimulation was relatively mild. The time course of this adaptation was rapid, and occurred within a fraction of a second, not after many hours like in the eye suture or sleep deprivation experiments.

Shew and colleagues (Shew et al. 2015) proposed a straightforward model to account for these results. They appealed to short term synaptic depression, which causes neurotransmission to become weaker after rapid, successive stimulations. This mechanism has been extensively studied and modeled and is well known to be prevalent in cortical synapses (Varela et al. 1997). Let us try to intuitively understand how this could account for their results. Before the onset of the movie, synapses are fresh and filled with neurotransmitter molecules. The first movie frames are then transmitted through the turtle visual system with very strong impulses, leading to amplified, or supercritical, propagation. After a few more frames and continued stimulation, many synapses become relatively depleted, causing the network to become less supercritical and to move toward the critical point. Tens to hundreds of milliseconds after stimulation, the depleted neurotransmitters are gradually restored. In this way, the network can balance the need for strong responses to important transients with the need to remain close to the critical point over longer time scales.

The model used by Shew and colleagues (Shew et al. 2015) built on a previous model proposed by Anna Levina, Michael Herrmann, and Theo Geisel (Levina, Herrmann, and Geisel 2007) that also incorporated short-term synaptic depression. There, Levina and colleagues argued that short term synaptic depression would be sufficient to organize cortical networks toward the critical point more generally, under conditions of continuous operation and not just during sensory transients. This is very closely related to the question of how cortical networks could wire themselves up during development to operate near the critical point. For answers to this question, let us turn to the next experiment.

Development toward the Critical Point

Neural development is the amazing process by which brains grow and connect, over a period of weeks to years, to become competent at surviving in their environments. Consider this process in humans, where some 86 billion neurons (Azevedo et al. 2009), each with 1,000 synaptic connections or more (Braitenberg and Schüz 1998), must wire themselves up so that an individual can speak a language, play tennis, learn to draw and sing, and navigate social situations. The information in the entire human genome is estimated to be about 750 megabytes, or about the contents of one compact disk (CD). The information required to assign all the connections from and to specific neurons is about 36×10^{14} bytes.[2]

To encode this, you would need over 4 billion CDs. Clearly, the information in the genome is vastly insufficient to specify the adult connectome.

How then are these connections properly directed to their targets? In addition to molecular gradients to steer gross connectivity of axons (Goodhill 2016), a large part of this process at the small scale is thought to be guided by simple rules like firing rate homeostasis, synaptic scaling, and spike timing-dependent plasticity that we mentioned earlier (Turrigiano and Nelson 2004). Coincident neural activity, as Donald Hebb predicted, goes a long way toward forming cortical maps that represent important features (Song and Abbott 2001). As a brain interacts with its environment, it gradually self-organizes to reflect the statistical structure of that environment so that it can encode stimuli, predict outcomes, and drive responses.

The trajectory of development across many species follows the same general pattern. At the beginning, there is an overabundance of neurons. These neurons then proliferate connections, far more than will be needed. The connections and neurons that are rarely used are then pruned. This massively trims the network, so that in humans, about a third of the neurons die after the first year of life (Hutchins and Barger 1998). Even during puberty, cortical thickness declines and this thinning process occurs more in highly intelligent children than in typical children (Shaw et al. 2006). Interestingly, this idea of trimming connections to improve performance has been adopted in machine learning, where they selectively cull the infrequently used connections (Molchanov et al. 2016).

As Ross Ashby, one of the pioneers of cybernetics, said, "a brain can improve till it fits its environment." Certainly, this is what neural development is thought to accomplish. It is natural to wonder if part of this improvement would involve coming closer to the critical point. This question has been addressed by several labs who have used model systems in culture to begin to find answers.

In one approach, Christian Tetzlaff and colleagues (Tetzlaff et al. 2010) isolated rat cortical neurons in suspension and then poured them out onto microelectrode arrays (MEAs), where they attached to the surface and began to sprout synaptic connections (figure 6.4A). This type of culture preparation, called "dissociated," or "primary," initially starts with very unstructured, almost random, patterns of connectivity. It takes about a week before neural firing starts, and then the cultures go through several stages over 3–4 weeks before a mature and stable firing pattern is established. Because these networks grow within a sealed culture dish, it is possible to record their activity many times over a period of weeks without breaking the sterile seal that is needed to keep them from becoming infected. These properties make this preparation useful as a simplified model system of neural development in cortical networks. Their spiking activity (figure 6.4B) produces neuronal avalanches across many electrodes (Pasquale et al. 2008), making it straightforward to plot avalanche size distributions across time.

When Tetzlaff and colleagues (Tetzlaff et al. 2010) did this, they found a familiar pattern. In the first two weeks, there was relatively mild activity which then greatly increased by the third week. This was followed in the fourth week by a period of reduced activity that stabilized by the sixth week. To track how close these cultures were to the critical point, they measured how much the avalanche size distributions deviated from an ideal power law; they called this measure Δp (figure 6.4C). The Δp measure was positive when the avalanche size distributions had a bump at the end, indicating supercriticality. It was

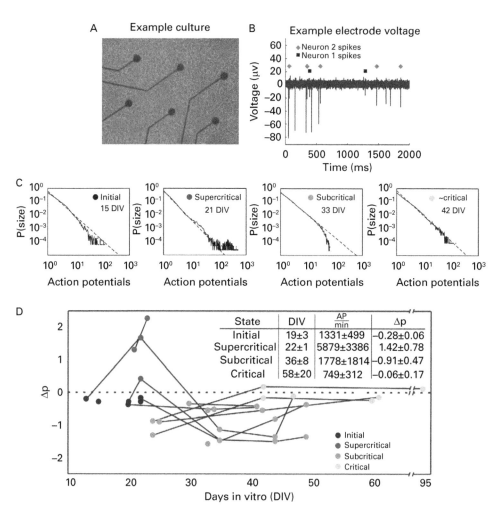

Figure 6.4
Cultures of cortical neurons develop toward the critical point. *A*, Example of a dissociated culture grown on a microelectrode array. Electrodes are black circles, 30 μm in diameter, connected by wires (straight black lines). Neuron cell bodies are irregular shapes, smaller than the electrodes, covering the glass surface of the array. *B*, Representative voltage traces from one electrode, showing how activity from individual neurons can be distinguished by action potential features. Panels A and B adapted from Timme et al. (2016). *C*, Representative avalanche size distributions plotted for four cultures at different stages of development, indicated by days in vitro (DIV: 15, 21, 33, 42). Dashed diagonal line shows expected distribution for an ideal power law. *D*, Summary plot of deviation from ideal power law, measured by Δp (y-axis), for cultures over many days (x-axis). Positive values of Δp indicate activity that is supercritical, negative values indicate subcritical, and near zero values indicate approximately critical. Each dot represents a culture and lines linking dots show cultures that were measured multiple times. Table shown in inset gives average values of DIV, action potentials per minute (AP/min) and Δp for the four stages observed. General trend was for cultures to start out slightly subcritical, then become supercritical, followed by a more strongly subcritical stage before approaching the critical point from below. The cultures that were closest to the critical point were those that were recorded at the latest periods. Panels C and D adapted from Tetzlaff et al. (2010).

negative when the distributions curved downward, indicating subcriticality. When it was near zero, the distributions closely followed a straight power law, indicating proximity to the critical point. They found that the cultures started out slightly subcritical, then became supercritical, followed by another subcritical period. They then gradually approached the critical point from below, so that all the mature cultures were very close to the critical point and stayed there (figure 6.4D).

These results indicate that neural development, at least in dissociated cultures, does indeed move toward the critical point, a result that was later confirmed by Levina and Priesemann (2017). It is also interesting to note that the cultures approach the critical point from two different directions: early on from above, and then later from below. This shows that in one model system there is evidence of bidirectional regulation toward criticality. In the previous experiments we discussed, only movement from one direction was shown at a time: eye suture from below, sleep deprivation from above, sensory adaptation from above.

Even though these cultures were not exposed to sensory inputs from the environment, they still managed to self-organize near the critical point. This suggests that local mechanisms alone would be sufficient for this process. To try to understand how this could happen, Tetzlaff and colleagues (Tetzlaff et al. 2010) made a computational model. This model included dendritic connections that sprouted in proportion to a neuron's *inactivity* and axonal connections that sprouted in proportion to the neuron's *activity*; the model also included firing-rate homeostasis. With these mechanisms, the network began with unconnected neurons and no activity. From there, it then followed three stages of development—sprouting of connections that led to the supercritical phase, pruning of connections that led to the subcritical phase, and an equilibrating stage guided by firing-rate homeostasis that brought the network close to the critical point. This recapitulated the general pattern of development that we discussed and matched the density of synaptic connections over time observed in the cultures (low-high-low). They found that the final, nearly critical, phase of the model was sensitively controlled by the strength of the inhibitory connections. Recall that this result is similar to the findings of Ma and colleagues (Ma et al. 2019). They noticed that in the models that accurately followed the experimental results, the inhibitory neurons, more than the excitatory neurons, accomplished tuning toward criticality.

In another study, this time performed on cortical slice cultures, Craig Stewart and Dietmar Plenz recorded neuronal avalanches over a period of 40 days (Stewart and Plenz 2008). Like the Tetzlaff study (Tetzlaff et al. 2010), they also found that activity levels started out low, grew, and then declined. Interestingly, though, they reported that the cultures, on average, remained close to the critical point throughout their development. While this result is somewhat different from that reported by Tetzlaff and colleagues, it still points toward maintenance near the critical point despite large changes in firing rates.

Themes from Homeostasis Results

Now that we have reviewed several experiments on homeostasis toward the critical point, let us try to discern general themes. We have some evidence of *adjustment from different directions*: in eye suture, it came from the subcritical regime; in sleep deprivation and sensory adaptation it came from the supercritical regime; in development it oscillated from both sides. We have also seen *different time scales* of recovery and *different magnitudes* of perturbation. Development occurs over months, recovery from eye suture over days, recovery from sleep deprivation over one night, while sensory adaptation occurs over subsecond time scales. Larger perturbations come with longer time scales of recovery, just as smaller perturbations take less time to return toward the critical point.

This diversity of scales and times suggest there are different mechanisms operating in each of these cases, but the general theme is very consistent—the system returns toward the critical point after being pushed away from it. The diverse mechanisms implicated by models include firing rate homeostasis, synaptic scaling, and Hebbian-like synaptic plasticity, along with sprouting and pruning of connections. There may also be important differences in the roles played by excitatory and inhibitory neurons in recovery toward criticality.

Could these results be explained simply by random drift toward a mean? This seems unlikely for several reasons. First, there is a repeatable structure in the recovery from eye suture: distance from criticality increases first and recovers first, while firing rate adjustments begin later and take longer. If everything were random, this order would be reversed sometimes, but that was not observed. This structured sequence suggests a cascade of processes is at work. Second, the recovery after sleep deprivation was not just slightly toward critical, but consistently overshot into the subcritical regime. If the process were random drift, after one night's sleep we would expect to see about half the subjects supercritical and about half subcritical; this was not observed. The fact that most subjects were well into the subcritical regime points toward a restorative process that was strongly activated by extended sleep deprivation. Third, recovery toward criticality in the sensory adaptation experiments always occurred with a rapid return that had no overshoot. If this process were a random walk, you would expect it to show varying times to get back near critical, some quick and some slow, and overshoot about half the time; none of the data exhibited this. Fourth, the approach toward the critical point in development followed a characteristic trajectory of supercritical, subcritical, then nearly critical. If this process were just random drift, then we would always expect to see roughly equal numbers of cultures above and below the time axis in figure 6.4D. Again, that is not what was observed, pointing toward a consistently structured cascade of mechanisms that bring the networks toward the critical point. While we do not yet know what the mechanisms are that bring the cortex toward the critical point and keep it there, it seems quite likely that they exist. Experimentally identifying these mechanisms, guided by hypotheses from modeling work (Hsu et al. 2007; Peng and Beggs 2013; Del Papa, Priesemann, and Triesch 2017; Girardi-Schappo et al. 2020), is expected to be an area of active research.

Health

If bringing the cortex near the critical point is actively regulated, then perhaps proximity to the critical point is important for neurological health. Homeostasis is common in the body, and has been observed for blood pressure, heart rate, breathing rate, body temperature and pH levels, to name but a few physiological indicators (Buchman 2002). When these homeostatic mechanisms fail, they can precipitate a crisis and even death. A corollary of this would be that departures from the critical point should be associated with neurological problems; there is now evidence that this is the case. The relationship between neurological disorders and departures from criticality is attracting increased research attention—for an excellent review, see Zimmern (2020). Rather than covering all the evidence, we will now focus on a few examples to illustrate this connection.

Seizures and Epilepsy

About 1 percent of all people will experience a seizure at some point during their lifetime (Begley and Durgin 2015). For those with epilepsy, seizures occur all too often—perhaps tens to hundreds of times a year. Seizures are characterized by excessive neural activity that interferes with movement, thinking, and memory. Because of its clinical importance, much research has been devoted to uncovering the etiology of epilepsy, but its exact causes are not yet known, and may differ between individuals (Ottman et al. 1996).

Here, let us examine the potential connection between excessive activity found in seizures, and runaway avalanches produced by the branching model. If seizures can be mapped on to the branching model, then we should expect epileptic tissue to have avalanche size distributions that depart from a linear power law by having an upward hump. We saw these humps before in the sensory adaptation experiments (figure 6.3F, G) and the development experiments (figure 6.4C). Do they appear during seizures?

Christian Meisel and colleagues (Meisel et al. 2012) examined seizures in humans from the framework of criticality and neuronal avalanches. To do this, they analyzed recordings from epilepsy patients who had surgically implanted electrocorticography (ECoG) arrays (figure 6.5A). These arrays are useful to surgeons who want to locate and remove the region of the brain that initiates the seizures, called the epileptic focus. The signals collected at ECoG electrodes reflect the summed activity of many neurons underneath each electrode, in the upper layers of cortex. While these signals do not show spiking activity in individual neurons, their amplitude and synchrony can give valuable information about neuronal populations. In fact, epileptologists often use ECoG synchrony to identify seizures (figure 6.5B, ictal period). Because these signals do not have spikes from individual neurons or sharply negative LFP signals, Meisel and colleagues (Meisel et al. 2012) instead used measures of synchrony called phase-locked intervals (PLIs) to assess proximity to the critical point.

To understand this, it will be necessary to take a digression into the Kuramoto model. Briefly, this model consists of many units that oscillate at the same frequency. You can imagine these units to be like mechanical clocks, each with its own pendulum. As Christian Huygens discovered back in 1665 (Ramirez and Nijmeijer 2020), when two such clocks are put next to each other on a shelf, thus sharing a physical connection, they eventually synchronize so that their pendula are moving in opposite directions. In this case, their phases differ by 180°. In a related manner, the connection strengths between units in the Kuramoto model control their synchrony, but here when the connections are strong the units will oscillate with a phase difference of 0°, demonstrating what is called phase locking.

The strength of the connections determines whether the network will be ordered or disordered. When the connection strengths are zero in the Kuramoto model, the units are independent and unsynchronized, with only random intervals where they happen to lock phases. In contrast, when all the connections are strong between units, they are highly synchronized, with nearly zero phase lags. This produces one long interval of phase locking, including all units. At the critical coupling strength between these two extremes,

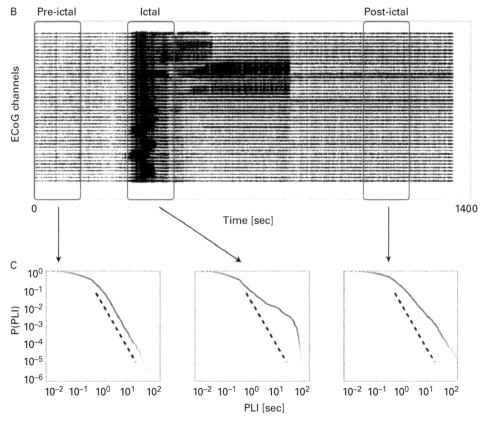

Figure 6.5
Seizures push the cortex away from the critical point. *A*, Example of an electrocorticography (ECoG) array placed on the cortical surface over a presumed epileptic focus in a human epilepsy patient. Adapted from Hobbs, Smith, and Beggs (2010). *B*, Voltage traces from ECoG. Rectangles indicate time periods before, during and after seizure (ictal) event; adapted from Meisel et al. (2012). Note large synchronous voltage fluctuations that characterize the seizure. *C*, Distribution of phase-locking intervals (PLIs), which have been used to identify dynamics near the critical point in human brain activity (Kitzbichler et al. 2009), plotted for the three time periods. For reference, dashed line shows the slope of a power law with exponent of −3.1. Distribution of PLIs in pre-ictal period is close to the dashed line and has an approximately power-law tail. During the ictal period, the distribution departs from the dashed line, showing that large PLIs occur significantly more often, indicating supercritical activity. The post-ictal period shows some movement back toward dashed line.

units will show a broad distribution of phase-locking intervals that follow a power law. A given unit may pop in and out of synchrony with any number of other units, for intervals at all scales. Thus, the Kuramoto model has a phase transition and a critical point; it was first used by Kitzbichler and colleagues to show that human fMRI and MEG data have a power-law distribution of PLIs (Kitzbichler et al. 2009).

With this background, we can now better understand Meisel and colleagues' findings. Before a seizure, the distribution of PLIs followed a power law (figure 6.5C, left panel, pre-ictal). During a seizure, this distribution rose above the previous power-law shape, indicating that the probability of longer PLIs increased (figure 6.5C, middle panel, ictal). After a seizure, this distribution moved back toward its previous power-law shape, but remained elevated (figure 6.5C, right panel, post-ictal). All eight subjects from this study showed a similar pattern (Meisel et al. 2012).

In another relevant study, Jon Hobbs, Jodi Smith, and I measured the branching ratio in tissue removed from juvenile epilepsy patients to reduce seizures (Hobbs, Smith, and Beggs 2010). This tissue was sliced and then put on 60 channel microelectrode arrays so that negative LFP signals could be detected. In four slices from different people, the tissue showed intermittent elevated activity periods where firing rates doubled or tripled (figure 6.6A). During these periods, the branching ratio showed a significant positive correlation with the firing rate, suggesting a positive feedback loop (figure 6.6B). This correlation was not seen during nonelevated firing periods, where the branching ratio hovered near 1 (figure 6.6C). In addition, the avalanche size distributions showed a hump during elevated firing periods, but not during nonelevated periods. These findings are consistent with the idea that seizures can be associated with supercritical activity (Hsu et al. 2007; Hsu et al. 2008).

Related to this, Christian Meisel found that when antiepileptic drugs were given to human patients with epilepsy, they significantly reduced the number of large avalanches and the strength of autocorrelations (Meisel 2020). He interpreted his results within the framework of a branching process, which would predict these things if antiepileptic drugs moved epileptic networks in the direction of subcriticality. Earlier work by him and colleagues showed that such drugs reduced measures of excitability in a dose-dependent manner (Meisel et al. 2015). Work by Ashrit Arviv, Oren Shriki, and colleagues (Arviv et al.

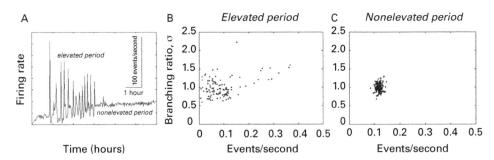

Figure 6.6
Supercritical branching ratio in tissue from juvenile epilepsy patients. *A*, Examples of elevated and nonelevated firing periods. This tissue exhibited a period of elevated firing 1 hour into the recording that was sustained for nearly 3 hours until it resumed a "normal" firing rate again for several more hours. All human tissue samples (*n*=4) exhibited these two distinct regimes. *B*, During elevated activity, there was a positive and significant correlation between the firing rate and the branching ratio, $p < 0.05$. *C*, During nonelevated activity, there was no significant correlation. Adapted from Hobbs, Smith, and Beggs (2010).

2016) also found upward deviations in avalanche size distributions during interictal activity, shortly after a seizure.

Taken together, these papers (Hobbs, Smith, and Beggs 2010; Meisel et al. 2012; Meisel et al. 2015; Arviv et al. 2016; Meisel 2020) are consistent with the idea that seizures occur when networks are supercritical. There, they have avalanche size distributions with positive humps and branching ratios that exceed 1. However, not all labs agree (e.g., see Hagemann et al. [2021]), and research in this area is still ongoing.

Bursting after Hypoxia in Infants

Let us turn now to an area where even slight changes in indicators of criticality have been developed to predict health outcomes. During birth, infants may unfortunately experience periods of reduced oxygen, known as hypoxia, and this can sometimes cause lasting brain damage. It is therefore crucially important to monitor brain activity right after hypoxic insult to assess the need for medical interventions, like cooling. Monitoring can be done by placing EEG electrode caps on infants. Right after hypoxia, EEG activity is suppressed, but later a bursting pattern typically develops (figure 6.7A). Previous attempts to use the mean or interburst intervals of this signal to predict outcomes met with limited success (Toet et al. 1999).

To improve this situation, Kartik Iyer and colleagues analyzed the data using tools developed from the study of critical phenomena (Iyer et al. 2014; Roberts et al. 2014). They squared the amplitude of each burst to obtain only positive profiles (figure 6.7A). These rectified bursts that crossed a threshold could then be quantified by their sizes and durations,

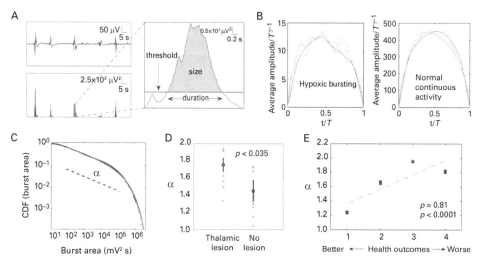

Figure 6.7
Signatures of criticality predict health outcomes in hypoxic infants. *A*, EEG trace from a human infant is squared so that burst sizes and durations can be extracted. *B*, Average burst shapes for infants after hypoxia (left panel) and for infants with normal, continuous activity (right panel). Note that hypoxic curves are skewed while normal curves are more symmetric. *C*, The cumulative distribution function for burst areas follows a power law with exponent α. Data are in black dots, while the power law fit with exponential cutoff is shown in gray. The power law spans four orders of magnitude. *D*, The exponent from activity several hours after hypoxia predicts which infants will develop thalamic lesions days later. *E*, The exponent measured several hours after hypoxic event significantly correlates with health outcomes measured days later. Panels A, C, D, E adapted from Iyer et al. (2014); panel B adapted from Roberts et al. (2014).

just like avalanches. They found that burst sizes followed a power-law distribution span-ning four orders of magnitude (figure 6.7C), which is notable because it is about two orders of magnitude larger than typically reported in microelectrode array studies. Statisti-cal tests showed that easily the best fit for these distributions were power laws with expo-nential tails. In addition, the average burst shapes were fractal copies of each other, and could be scaled to show a reasonably good collapse (figure 6.7B). Finally, all the exponents satisfied the exponent relation, showing that this system was poised near the critical point.

Infants who recovered well from hypoxia had avalanche size exponents that were sig-nificantly smaller than the exponents from infants who had poor health outcomes (fig-ure 6.7D, E). In other words, the healthy infants had a broader distribution of avalanche sizes, even though both groups had scale-free distributions. The poor health outcomes included thalamic lesions as assessed by structural neuroimaging, and more developmental abnormalities, as rated on a scale (1 = normal; 2 = mildly abnormal; 3 = moderately abnor-mal; 4 = severely abnormal). Moreover, the infants with poor outcomes produced avalanche shapes that were not symmetric, but positively skewed (figure 6.7B). The authors specu-lated that this positive skew was caused by reduced inhibition at the start of bursts, making them rise faster than the symmetric bursts.

While these indicators are still within the realm of critical, they are perhaps moving in the direction away from it. For example, a power law with exponents larger than 2, while still scale-free, indicates a system where correlations do not extend infinitely (see the appendix). Likewise, asymmetric burst shapes suggest an imbalance between excitation and inhibition that, if continued, would lead to supercritical activity. In any case, these minor deviations are still significantly predictive of poor health outcomes and are an advance over previous methods. This approach has now been extended successfully to predict outcomes in extremely preterm infants who are 22–28 weeks old (Iyer et al. 2015).

Now that we have looked at how pathological conditions may be related to departures from the critical point, we can next see how healthy brains have signatures of being close to the critical point in both behavior and neural activity.

Relating Behavior to Criticality in the Brain

Psychologists have noted for over a century that human perceptual behavior can be char-acterized by scaling laws that approximately follow straight lines in log-log plots (Kello et al. 2010). The underlying causes of these laws remained mysterious for over 100 years. Recent work related to scale-free activity in the brain may have opened a window on the origins of this perceptual behavior. We will now describe these scaling laws in more detail and the experiments that have shed light on their causes.

We are usually quite good at noticing slight differences in our surroundings. For example, if I were to hand you two approximately 100 g masses, you could probably tell if they dif-fered by 5 g. This is what we would call a just noticeable difference, or JND. However, you would probably not be able to tell a difference if I gave you a 1,000 g mass and a 1,005 g mass. It turns out that our ability to discriminate between two masses is proportional to the size of the masses. As the masses grow, the JND grows with them. If we call the mass M, and the JND of the mass ΔM, then the ratio of $\Delta M/M$ is approximately constant over many scales—indicating scale-free behavior. Notably, scale-free behavior is true for noticing dif-ferences in light or sound intensity as well as changes in sugar concentration; this appears to be a law of perception for all sensory modalities. Because Weber and Fechner were among the first to quantify this relationship, this is often called the Weber-Fechner law of psychophysical perception (Weber 1996).

But these scale-free relationships are not confined to sensory systems. They also turn up in behavioral processes like reaction times, memory recall, and foraging for resources (Kello et al. 2010). It is unclear why these all follow scaling laws. One of the hypotheses is that they reflect the nearly critical dynamics of the cortex (Kello 2013).

Matias Palva, Satu Palva, and colleagues were interested in exploring the possible relationship between behavioral performance and markers of criticality (Palva et al. 2013). To assess behavioral performance, they chose a very simple task. Human subjects were asked to detect auditory and visual stimuli against a background of noise, where stimuli were presented hundreds of times (figure 6.8A). Before the testing run, saliences were adjusted so that subjects could detect stimuli only about 50 percent of the time. It had been well documented that in tasks like these, successful identifications were not randomly spread across time. Rather, they tended to show slow fluctuations lasting 1 to 1,000 seconds, where, for example, subjects would show a patch of many successes, followed by a patch of many failures (figure 6.8C, upper two rows). Over long periods, these fluctuations had fractal structure and power-law scaling in time. This structure can be quantified by measuring long-range temporal correlations (LRTCs); these are not found in completely random sequences of successes and failures (Teich et al. 1997; Hardstone et al. 2012; Ihlen 2012). (See the appendix for a more detailed description of LRTCs and how to measure them.)

Palva and colleagues knew from their previous work (Linkenkaer-Hansen et al. 2001) that brain activity, as measured by EEG and MEG, also showed significant LRTCs. For example, enhanced oscillations near 10 Hz did not occur at random times but tended to appear in clusters lasting up to a thousand seconds with fractal structure. In addition, they knew that others had used MEG signals to extract neuronal avalanches from human subjects (Shriki et al. 2013). As we have discussed, these avalanches occur at much shorter time scales, and typically last less than half a second. Despite these tremendous differences in time scales, they hypothesized that the scaling exponents in all three domains (behavioral performance, LRTCs, neuronal avalanches) would be significantly correlated. If correct, this hypothesis would reveal a potential connection between neuronal avalanches and behavior in healthy humans.

Their results confirmed that the scaling exponents in behavior were significantly correlated with those from the LRTCs of brain activity (figure 6.8G). In addition, they showed that the exponents from neuronal avalanches were significantly correlated to those from LRTCs from activity and from behavior (figure 6.8H). Moreover, they were able to identify the brain areas that were related to these long-term correlations (figure 6.8B). For the visual stimuli, this included visual areas and attentional areas. Similarly, for the auditory stimuli, this included brain areas implicated in hearing and attention. This result is important because it shows the correlations were generated by brain regions performing the behavioral tasks and rules out the possibility that all these correlations were merely some epiphenomenon not related to neuronal function.

This important paper unified two largely independent lines of work by showing that neuronal avalanches at the microscale were related to the macroscale fluctuations that occurred across the entire cortex. This paper also strengthened the connection between

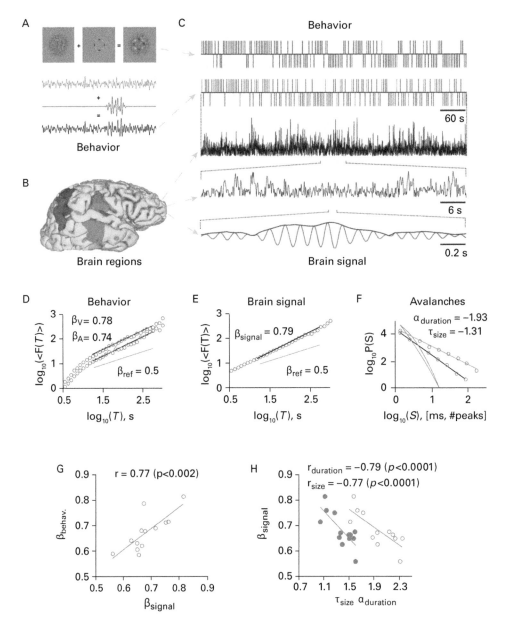

Figure 6.8
Exponents from behavior and brain signals are significantly correlated. *A*, There were two behavioral tasks. The upper row shows a random visual cue with a signal that was sometimes added. The lower 3 rows show random noise with a chirp that was sometimes added. Subjects were to detect signals embedded in noise; their success rate was titrated to be near 50 percent. *B*, Cortical regions involved in the tasks, highlighted in darker gray, correspond to audio and visual processing. *C*, The upper two rows show results of behavioral tasks. Correct trials are upward ticks and incorrect trials are downward ticks; visual results are above auditory results. The lower three rows show brain signals at progressively shorter timescales. *D*, Correct trials have long-range temporal correlations such that the variance on the y-axis scales with the length of the time window on the x-axis. The slope of this relationship is given by the exponents β_V and β_A for visual and auditory tasks, respectively. The exponent for a random walk is $\beta_{ref} = 0.5$ and is what would be expected by chance. *E*, Brain signals also show long-range temporal correlations. *F*, Neuronal avalanches from brain signals were better fit by power laws than other functions. Shuffled surrogate data, in gray curves, was not fit by power laws. *G*, Exponents from behaviors are significantly correlated with the exponent from brain signals in each subject. *H*, Exponents from brain signals are significantly correlated with exponents for avalanche sizes and durations. Adapted from Palva et al. (2013).

neuropathologies and the framework of criticality through LRTCs. For example, Tessa Montez, also working with Klaus Linkenkaer-Hansen and others, found that early-stage Alzheimer's patients had reduced exponents in long-term temporal correlations (Montez et al. 2009). Christian Meisel and colleagues found that extended sleep deprivation reduces these exponents as well (Meisel et al. 2017). Both these findings make sense intuitively— as the cortex moves away from the critical point, temporal correlations should decrease. It also sets the stage for new research to examine neuronal avalanches explicitly in diseased populations. While this work remains to be done, a solid scientific foundation for it has already been established.

Chapter Summary

If operating near the critical point is indeed useful to the cortex, then one would expect to find evidence that it develops toward the critical point as it matures, and that it returns there after being perturbed away from it. In this chapter, we reviewed several experiments showing that development and homeostasis cause cortical networks to move toward criticality. We also saw that neuropathologies can be characterized by departures from the critical point. Complementary to this, healthy human behavior can be characterized by scaling laws. We reviewed research that shows the exponents from behavior are significantly correlated with exponents from long-range temporal correlations and exponents from neuronal avalanches. Taken together, the findings reviewed in this chapter suggest that neurological and behavioral health is improved when the cortex operates near the critical point. From this perspective, it makes sense that there are processes to drive the cortex toward the critical point and return it there after disruptions. These processes in the brain are just a natural extension of the homeostatic functions Walter Bradford Cannon envisioned for the body about a century ago.

Exercises for this chapter can be found through a link given in the appendix.

7

Quasicriticality

In science it often happens that scientists say, "You know that's a really good argument; my position is mistaken," and then they would actually change their minds and you never hear that old view from them again. They really do it. It doesn't happen as often as it should, because scientists are human and change is sometimes painful. But it happens every day.
—Carl Sagan

When I first began to work in the area of criticality in the brain, I was thrilled to see our data follow power laws, the branching ratio to be near 1, and to realize from a simple model that information transmission would be optimized near the critical point. I was deeply in love with the hypothesis that the brain was critical, and often went to sleep dreaming about it. As the years passed, though, our relationship became strained. There was controversy and data from others that made me at times start to doubt its validity. The most jarring news, though, came from my trusted colleague, Gerardo Ortiz, a condensed matter theorist. He and our graduate students Rashid Williams-Garcia and Mark Moore, slowly and calmly approached me with arguments that the brain might not really be critical. Although I almost felt dirty as I listened to them, I was resolved to be a brave scientist and value logic and evidence above my emotions. Over time, and reasoning together with them, I slowly came to grips with a new truth. This chapter is about that truth.

While the above paragraph is intentionally melodramatic, it does capture something of the feelings I had upon realizing that the cortex can only operate *near* the critical point, and not exactly *at* it. If the cortex cannot be critical, then what is it? The answer is that it is quasicritical. As I will explain, this means it is operating close enough to the critical point to enjoy the benefits of optimality and close enough to appear to be universal. As we will see, something keeps it from landing right on the critical point. That something also alters the picture of perfect universality.

Having just told you in chapter 5 about the wonders of universality, I am embarrassed to now admit that the story is incomplete. While there was much good evidence in support of universality presented there, as we noted at the end of the chapter it was not wrapped up yet. We have entered one of the currently open questions in the field; a major part of this chapter will be devoted to explaining the problem more clearly and attempting to resolve it. To frame it as concisely as possible, universality demands that there should be only one set of characteristic exponents (α, τ, γ), yet the data in the literature only show a consistent value for γ in spiking cortical networks. The other exponents, α and τ, while constrained by the exponent relation, do not assume constant values but vary over a range.

But isn't this a minute technical problem? Why not just say "close enough" and move on? The reason this is worth fighting for is that universality is central to our ability to simplify complex systems and transcend details. As I said before, many physicists consider the discovery and understanding of universality to be one of the most profound accomplishments of physics, if not science. If it is true, then universality conceptually overthrows the causal dominance of lower levels; it shows that emergent phenomena can stand on their own at larger scales. If it applies to the cortex, it would be revolutionary. But it can only apply if there is one set of exponents for the cortex, not several.

In this chapter, I will first describe the current difficulties with universality in more detail. I will then offer a potential solution, based on a new principle for brain dynamics called "quasicriticality." As we will see, there are both simulations and data consistent with quasicriticality's predictions. We will then look at alternative explanations for how cortical networks operate in the vicinity of the critical point. Continuing with the theme of the previous chapter on homeostasis, we are here covering relatively new material, on the frontiers of our knowledge. Regardless of which of these ideas turns out to be correct, they all share in common that the cortex cannot be operating exactly at the critical point. While that was for me a disappointment, it appears to be the truth.

Universality: Unfinished Issues

To get the broadest view possible on universality, let us go back and think about the different ways that critical exponents from the data could have turned out. First, it could have been the case that each species had its own set of critical exponents. Rats, mice, monkeys, and turtles would all be different points on the plane, and these points could even change depending on whether the animal was awake or under anesthesia (figure 7.1A). Here, everything would be particular and overall, nothing would be universal. In contrast, perfect universality would appear as a single set of critical exponents, on the exponent relation line, for all species, under all conditions. This is represented by the single point in figure 7.1B. An intermediate scenario would be for each species to have its own set of exponents, all adhering to the exponent relation (figure 7.1C). This would argue that each species has its own universality class.

Instead of these situations, the one we have is far more puzzling. We see exponents for many different species, awake and stimulated or under anesthesia, all adhering to the exponent relation line. Since there is not a single dot, there cannot be universality. But what is most peculiar is that even a given individual rodent can move along this line at different times of day (figure 7.1D). Because the points are on the exponent relation line, that rat must in some sense be operating near the critical point.

But to think that a single rat might change universality classes throughout the day seems preposterous. The whole idea of universality is that there is something general overriding the details. Under this hypothesis of multiple universality classes for an individual, the details would be dominating the general, so it would make no sense to talk about universality

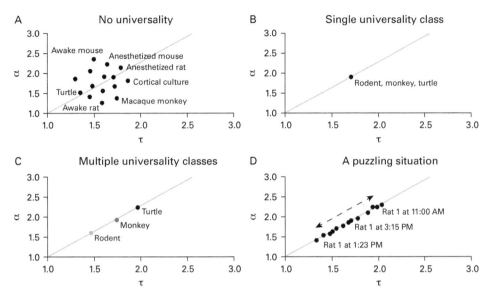

Figure 7.1
Four possible scenarios with universality. Each black circle represents the avalanche duration (α) and size (τ) exponents from one dataset. The thin gray line represents the exponent relation found in Fontenele et al. (2019), previously shown in figure 5.3. *A*, The case of no universality. Each different species and condition produces a separate pair of exponents, and most of them avoid the exponent relation line. *B*, Universality would occur if all species under all conditions had the same set of exponents, and these adhered to the exponent relation. *C*, Multiple universality classes could occur if, for example, each species had its own set of exponents and they all obeyed the exponent relation. *D*, The actual situation is far more puzzling. A given individual can have different exponents at different times or under different conditions, and these exponents nearly satisfy the exponent relation. The data are thus close to critical, but not universal. This pattern of lying along the line is not only true for one individual, but for many species, whether under anesthesia or awake. These plots are not taken from actual data, but schematically represent the conceptual differences between hypotheses. Adapted from Fosque et al. (2021).

at all. But this scenario is not hypothetical—this is exactly what was published in Fontenele et al. (2019) (see also figure 5.3), and what we have found with our own data in Fosque et al. (2021) (see also figure 3.12). As the environmental inputs were changing, the exponents were moving up and down along the exponent relation line. Reconciling this with universality is challenging.

A Possible Solution: Quasicriticality

What could make the data move along the line? One candidate would be the external drive that is being supplied to a given patch of cortex. Recordings from single cortical neurons in vivo show us that they are constantly receiving synaptic inputs, even when an animal is anesthetized (Doi et al. 2007; Sun and Dan 2009). A conservative estimate would be that these synaptic currents arrive at a rate of 5 Hz or more. Given this, a cubic millimeter of cortex containing 50,000 neurons must be receiving a quarter of a million inputs per second; it is clearly not in equilibrium.[1] Sudden changes in sensory inputs could easily make external drive to sensory cortices increase by several factors of 10. Perhaps changes in external drive could cause critical exponents to move along the exponent line.

To see how this might work, let us again go back to the branching model. There, the rate at which activity is initiated is controlled by the parameter P_{spont}. This gives the probability that a neuron will become spontaneously active. When $P_{\text{spont}} = 0$, there is no activity in the model and there are no avalanches. Clearly, we must have $P_{\text{spont}} > 0$ for anything to happen but making P_{spont} too large could cause random activity to obscure the cascading activity of the avalanches. Under those conditions, we would not have power laws either. The ideal level of spontaneous activity therefore would be large enough to trigger avalanches but small enough to rarely interfere with avalanches once they were triggered.

This issue came up decades ago in the field of self-organized criticality (SOC) (Jensen 1998; Marković and Gros 2014). There, the chosen model system involved avalanches in sandpiles (Bak, Tang, and Wiesenfeld 1987). To trigger an avalanche, a single grain of sand was dropped onto the top of the conical sand pile. The time between new grains of sand being added had to be much longer than the time it took for an avalanche to travel through the sandpile, so that a new grain would never be added while an avalanche was happening. This separation of timescales ($T_{\text{sand}} \gg T_{\text{avalanche}}$) was so important that the early researchers in SOC insisted that it was a necessary condition for any system to self-organize to criticality (Bak 1996). In our system, this would be equivalent to a very low value of P_{spont}. If external drive is added to a SOC model in violation of this condition, we could have what Juan Bonachela and Miguel Muñoz call self-organized quasicriticality (SOqC) (Bonachela and Muñoz 2009). Such a situation is very realistic, as a separation of timescales does not seem to apply to a given region of cortex—it is always receiving inputs from other parts of the brain.

With this background, now let us consider how spontaneously generated avalanches in a small population of neurons would be altered if there were additional activation provided by axons coming from another area. These axons could excite a small fraction of neurons to fire when they otherwise would have remained silent; this would be like increasing P_{spont}. The effect of this external drive, at least on some occasions, would be to concatenate two previously independent avalanches (figure 7.2C), an effect that was previously noticed by Viola Priesemann and colleagues (Priesemann et al. 2014) and our group as well (Williams-Garcia et al. 2014). If this happened often enough, it would cause the avalanche size distribution to shift so that more large avalanches were represented. This would require a longer tail and would make the slope of the distribution in log-log space become shallower. Therefore, the exponent would be effectively[2] reduced (figure 7.2D). Notice that this should happen for both avalanche sizes and avalanche durations, as a chance concatenation would increase both (Helias 2021). Simulations of the branching model show this to be the case when P_{spont} increased (Fosque et al. 2021; Williams-Garcia et al. 2014).

Using this idea, if we were to plot how both exponents would change under increased P_{spont}, we would see them move along the exponent relation line downward and toward the left (figure 7.3). In contrast, reduced P_{spont} would cause the exponents to move upward to the right, and in the limit of $P_{\text{spont}} = 0$, they would stop moving, coming to rest at some theoretical ideal.

Under this picture, if there were a place to find the exponents for the universality class, it would be in the upper right of the plot. Paradoxically though, such a network would never be active. This means that the exact critical point would be practically unattainable in this model. The best way to approximate the critical point would be to run a simulation where P_{spont} was so small that it would only trigger avalanches and never interfere with them. This

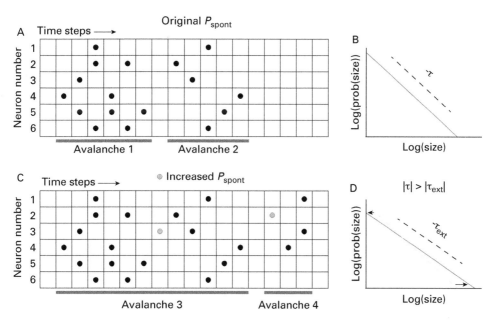

Figure 7.2
How increased external drive reduces the magnitude of exponents. *A*, A raster plot produced under low probability of spontaneous activity (P_{spont}). Two avalanches are shown; neuron firings are black circles. *B*, If the network is nearly critical, the avalanche size distribution will approximately follow a power law with exponent τ. *C*, The presence of external drive can be modeled by an increase in P_{spont}. This causes new activations (gray circles) that would not otherwise have occurred. This additional activity, though rare, can concatenate two avalanches that were previously separate (Avalanche 1, Avalanche 2) to produce a longer and larger avalanche (Avalanche 3). External drive can also trigger new avalanches (Avalanche 4). *D*, Increased avalanche sizes make the tail of the distribution move further along the x-axis, causing its slope to decrease. The corresponding effective exponent therefore decreases in magnitude, $|\tau| > |\tau_{ext}|$. Plots here are presented in schematic form for clarity but reflect results from simulations and experiments in Fosque et al. (2021) and Williams-Garcia et al. (2014).

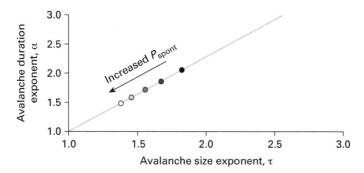

Figure 7.3
Moving along the exponent relation line. As the probability of spontaneous activity P_{spont} increases, the effective exponents (α, τ) decrease in magnitude. Because they continue to satisfy the exponent relation, they move along the line. In all cases, they are close to the critical point. If there is a universality class, it would exist toward the upper right, in the limit of zero spontaneous activity. Again, this is a schematic representation, for clarity, of results from Fosque et al. (2021).

can be artificially arranged by triggering an avalanche, then waiting until it has fully run its course before triggering another one. This is relatively easy to do in simulations, allowing us to plot what the phase diagram would look like in the theoretical limit of $P_{\text{spont}} = 0$.

This shows that increased external drive (rising P_{spont}) could explain why the exponents move along the line. It also preserves the idea of the universality class—it exists in the limit of $P_{\text{spont}} = 0$. Universal behavior can be approached closely when there is little drive. For this to be a more specific hypothesis with testable predictions, though, we will need to think more carefully about its other implications. To do this, let us examine related features of the system that will also be affected by increased external drive.

The phase diagram is one such feature. Before we consider changes to it, let us first recall how it looks without external drive. In figure 7.4A, we see an inactive phase on the left and an active phase on the right. These two phases have a transition when the branching ratio σ exactly equals 1. When $\sigma < 1$, even triggered activity will, on average, not lead to a subsequent avalanche. This causes the density of active sites, ρ, to be zero to the left of the phase transition point. Only when $\sigma \geq 1$, to the right of the phase transition point, will we get nonzero average activity in the network after a triggering event. And as σ is increased, this activity will grow with it. There are two clear phases and a critical point exactly at the transition between them, just as we saw previously in figure 3.3 for the branching model.

Now, when external drive is applied to the network, the inactive phase to the left of the critical point disappears because the network is always receiving inputs (figure 7.4B). This means that there is activity on both sides of $\sigma = 1$, and thus there is no longer an inactive-active phase transition. With this phase transition gone, there is no longer a critical point either. This is yet another reason the cortex cannot, in the strict sense, ever be actually critical. As long as there is any external drive, and the data tell us that this will always be the case, then cortex will not have a critical phase transition point. Interestingly, we have seen glimmers of this before in actual data (figure 3.10), where the presumed subcritical phase showed reduced but nonzero activity.

In addition to erasing the phase transition, increased external drive affects how the network reacts to inputs. Recall that the susceptibility measures how responsive the network will be to slight changes in inputs. At the critical point, in an infinite network, the susceptibility diverges to infinity. Intuitively, this means that activating one neuron in a critical network has the potential to unleash an avalanche of infinite size. Even if this does not happen often, the fact that it can happen causes the average network response to become infinite in critical networks.

However, this situation is different once external drive is introduced. Now, the effect of a single additional activation is much less dramatic. This is because the network may already be experiencing 100 external activations in 1 second—how much would things change when that number becomes 101? As a result, the susceptibility curves are blunted, becoming lower (figure 7.5C). In the extreme limit of large external drive, the intrinsic network response would be nearly undetectable.

There is yet another important consequence of increased external drive. This has to do with the value of the branching ratio σ at which the network begins to show increased activity. If there is no external drive, then the network will begin to show increased activity only when the branching ratio is greater than or equal to 1 ($\sigma \geq 1$). But with external drive, this rise in activity occurs even earlier, for values of σ less than 1. The size of the

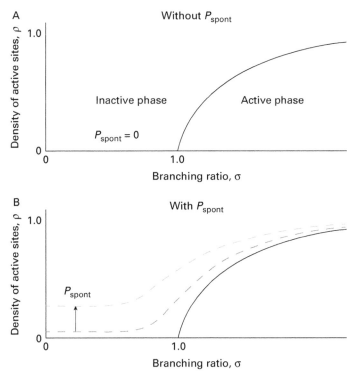

Figure 7.4
How external input changes the phase diagram. *A*, A schematic of the ideal phase diagram for the branching model in the limit of zero spontaneous firing probability ($P_{spont}=0$), showing an inactive phase on the left and an active phase on the right. The critical point occurs exactly at a branching ratio of 1. The order parameter is the density of active sites, ρ, and the control parameter is the branching ratio, σ. This is similar to figure 3.3. *B*, When external drive is applied (represented by increased P_{spont}, upward arrow), the inactive phase disappears and there is no phase transition. Dashed lines show how the density of active sites ρ would be nonzero even for values of the branching ratio σ below 1. Under these conditions, there is no critical point. More external drive causes the curves to rise further. Note that for larger values of P_{spont}, the dashed lines begin to curve upward well before the branching ratio equals 1. Figures are schematic and exaggerated for clarity but reflect actual results from simulations and data (Williams-Garcia et al. 2014; Fosque et al. 2021).

effect grows with increased external drive, as shown in figure 7.4B. It is as if the effects of being near the critical point are sensed much earlier when there is external drive. This causes the susceptibility curves to become broader, and they start to rise well before $\sigma = 1$.

We can visualize these effects better if we track the branching ratios at which the susceptibility curves peak. As external drive increases, these branching ratios get smaller. If we identify these branching ratios in the phase diagram (figure 7.5B), we find that a line can be drawn through them, connecting them to the ideal critical point. This line traces out where the maximum susceptibility can be found as the branching ratio σ, the control parameter in the model, is changed. This line of maximum dynamical[3] susceptibility also turns out to be the line along which mutual information is maximized (Williams-Garcia et al. 2014). Because related effects were predicted long ago in phase transitions of fluids by Professor Benjamin Widom,[4] this has been called the "Widom line" in the literature.

The organizing principle of quasicriticality predicts that when external drive is increased, the network will move in a region centered along the Widom line known as the quasicritical region. In doing so, it will continue to have optimal susceptibility and information

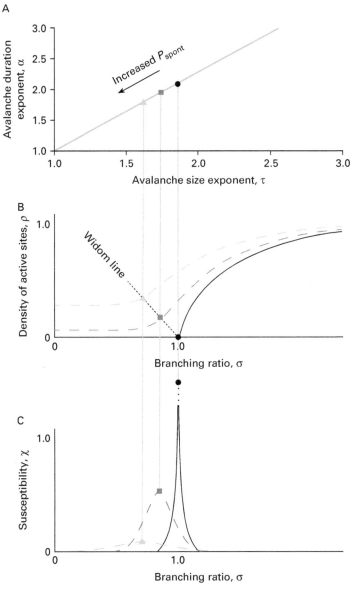

Figure 7.5
How moving along the line changes the branching ratio and the susceptibility. Here the relationship between all the plots is shown. *A*, As external drive is increased, exponents decrease (black to light gray). Circle is idealized universality class with zero external drive; square is mild external drive; triangle is further increased external drive. All exponents continue to adhere to the exponent relation, showing they are still close to the critical point. *B*, Increases in external drive cause the dashed lines to lift above zero, removing the phase transition. Black circle represents ideal critical point with no external drive. The square and triangle are at places of maximum curvature for their respective levels of external drive. These points are analogous to the critical point, but with external drive; they are *quasi*critical points. We expect to see peaks in susceptibility, though not infinite, at these points. The line connecting these quasicritical points to the critical point is called the dynamical Widom line. Note that it tilts to the left, predicting that branching ratios at which susceptibility is maximized will decrease under external drive. *C*, Susceptibility curves for the three points. For the black circle, a true critical point, the susceptibility will diverge to infinity. As external drive is increased, susceptibility curves become lower and broader. Their peaks correspond to points along the Widom line, and occur at lower values of the branching ratio σ. These are schematic representations, where distances are exaggerated for clarity, of results from Williams-Garcia et al. (2014) and Fosque et al. (2021).

transmission for that condition (Williams-Garcia et al. 2014). While the network cannot be critical in the quasicritical region, it will still be close to the critical point and therefore will still approximately adhere to the exponent relation. From this perspective, it is as critical as possible (Helias 2021).

There are several testable predictions from this idea. First, as external drive is increased, the height of the susceptibility curve will be decreased. Second, as external drive is increased, the location of the peak of the susceptibility curve will occur at decreased branching ratios. A third prediction, but one that was already present in the data, is that the exponents will move downward and to the left along the exponent relation line as external drive is increased. These predictions all follow from observing how simulations of the branching model behave as external drive is increased.

These predictions have been tested on datasets from cortical slice cultures. There, natural fluctuations occur in the branching ratio, susceptibility, and firing rate over a 1-hour recording; when the recording is broken up into many smaller time windows, curves can be plotted for all these variables. Although only three datasets contained enough variability near the critical point for us to plot these curves, we found their results to be consistent with the predictions of quasicriticality (Fosque et al. 2021). More stringent tests in the future should look at causally manipulating external drive through electrical or chemical stimulation.

While the results look promising for quasicriticality so far, this is still a relatively new research area and far from settled. There are other competing ideas that could be found to match the data. Let us now turn to those and review their predictions for how neural networks operate in the vicinity of the critical point.

Another View: Slightly Subcritical

Viola Priesemann and colleagues (Priesemann et al. 2014) have also concluded that the cortex cannot be critical; they have argued that it operates in a slightly subcritical regime. There, it is far enough away from the critical point to be safe from seizures, yet close enough to enjoy nearly optimal information processing. This would seem to be the best of both worlds (figure 7.6).

To arrive at their position, they analyzed spike recordings in vivo from awake rats and monkeys and anesthetized cats, as well as LFP recordings from humans (Priesemann et al. 2014). They compared these data to the output of simulations that could be tuned around the critical point. The best match between their simulations and the data occurred when three conditions were met. First, the output from the simulations had to be subsampled, meaning that the activity of some neurons at some times would be missed. This is a very realistic assumption, as even the densest electrode arrays invariably fail to record from some neurons. Second, the simulations had to be driven by external inputs. Again, this is quite reasonable and agrees with what we mentioned earlier about constant drive to cortical areas. Third, they had to tune the simulations to be slightly subcritical; instead of a branching ratio of 1.0, they found that one of 0.99 fit the data better. This is consistent with the idea that the cortex is not conservative, meaning that whatever input is delivered to it will eventually die out over time because it is not perfectly preserved as it travels through multiple synaptic stages. Other authors have also used this point to conclude that perfect criticality should not be possible in the cortex (Bonachela et al. 2010).

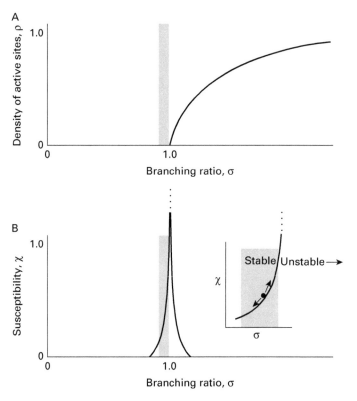

Figure 7.6
Slightly subcritical. *A*, Phase diagram of a system with an inactive-active phase transition. Slightly subcritical networks would operate in the gray zone. *B*, Susceptibility curve with corresponding gray zone. Inset: Expanded view of stable and unstable regions of the susceptibility curve. The network could travel up the curve as σ is increased, improving information processing. Traveling beyond the critical point, though, would lead to run-away avalanches and instability.

The slightly subcritical position was further bolstered a few years later when Jens Wilting and Viola Priesemann developed a far more accurate method for estimating the branching ratio from sparsely sampled data (Wilting and Priesemann 2018). When this was applied to a wide range of datasets, the results were consistent and showed them to be slightly subcritical, with branching ratios typically in the range of 0.99 to 0.95. Interestingly, when they applied their methods to spike data from epilepsy patients, they found that the branching ratio remained below 1 even before and during seizures (Hagemann et al. 2021).

Viola Priesemann and her group have more recently started to refer to this slightly subcritical state as the "reverberating regime" (Wilting and Priesemann 2019b). This indicates that a transient stimulation delivered there does not die out swiftly, as would be expected when the network is subcritical; it does not die out with a power-law tail either, as would be expected at the critical point. Rather, it dies out slowly, causing reverberations or echoes that last for a while.

In some respects, the slightly subcritical/reverberating viewpoint is similar to that of quasicriticality. Both note that external drive is important and predict that branching ratios will be below 1. They also posit that the cortex will generally be close to the critical point. They differ in that quasicriticality makes specific predictions about how increased external drive will decrease the branching ratio, susceptibility, and information transmission (Williams-Garcia et al. 2014). In addition, quasicriticality makes specific predictions about how the effective exponents will move along the exponent relation line (Fosque et al. 2021). These different predictions could be tested experimentally to help us distinguish between the two positions.

Another View: Subsampling

A very different explanation for why the effective exponents move along the line has been offered by Tawan Carvalho, Mauro Copelli, and colleagues (Carvalho et al. 2020). They investigated how the fraction of neurons sampled in a population might affect estimation of the effective exponents. They varied the sampling fraction from 100 percent all the way down to 0.5 percent in a simulated neural network containing 100,000 neurons. Each neuron was connected to all the others, giving the model all-to-all connectivity. When fully sampled, the model produced exponents that were very close to those expected for the branching model with all-to-all connectivity ($\tau = 1.5$, $\alpha = 2$, $\gamma = 2$), which is in the universality class of mean field-directed percolation (see squares in figure 7.7). This is clearly off the exponent relation line for $\gamma = 1.3$ that is typical of cortical spike data, but when they sampled fewer neurons, they found the exponents began to converge onto that line, moving upward and to the right (see triangles in figure 7.7).

Why might this happen? When all neurons are sampled, an avalanche traveling through the network can be completely tracked without interruption; large avalanches will be accurately accounted for. But this situation is very different under subsampling, where many neurons are omitted. These omissions can introduce gaps within a large avalanche, making it appear as if it were several smaller avalanches. Note that this is the reverse of what

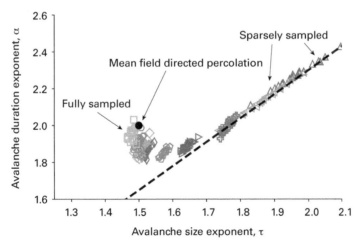

Figure 7.7
Decreased sampling can move effective exponents along the line. The dashed line indicates exponents that satisfy the exponent relation when $\gamma = 1.30$. When output of the simulation is fully sampled (squares), results are very close to the mean field for directed percolation (black circle). As the fraction of neurons sampled decreases (diamonds, pentagons, plus signs, triangles), the effective exponents increase and begin to converge onto the dashed line. Adapted from Carvalho et al. (2020).

happens when P_{spont} is increased in quasicriticality—instead of concatenating avalanches, subsampling splits them. With fewer large avalanches, the tail of the avalanche size distribution moves to the left, making the slope steeper and causing the effective exponent to grow. As the effective exponents increase, they climb upward and to the right in the plot.

This view generates several testable predictions. Those data points in the upper right of the exponent plot should generally have fewer numbers of neurons in their recordings, while those in the lower left should have more neurons. Any movement along the line by an individual should be matched by changing numbers of neurons in the recordings. Another prediction that could be tested on larger recordings, with several hundred neurons, would be to see if subsampling these datasets down to 100 or fewer neurons generally increases the effective exponents, causing them to move upward and to the right.

Another View: Griffiths Phase

Before closing this section, it will be helpful to consider one more view on what is happening very near the critical point. This position is based on the fact that network structure can strongly affect the dynamics of a neural network. While this general point has been known for some time (Van Vreeswijk and Sompolinsky 1996; Netoff et al. 2004; Rajan and Abbott 2006), the way network structure influences dynamics specifically near the critical point has only been appreciated more recently in neuroscience.

Many cortical simulations are conducted on randomly connected networks (figure 7.8B); these are relatively easy to construct and analyze. But we know that living neural networks have many highly nonrandom features (Song et al. 2005). For example, at the whole brain level there are many modules for each sensory modality and there is a hierarchy of modules (Felleman and Van Essen 1991; Siegle et al. 2021). At the microcircuit level, there is evidence for highly connected hub neurons (Bonifazi et al. 2009; Shimono and Beggs 2015), and even denser subnetworks of hub neurons within the broader network (Dann et al. 2016; Nigam et al. 2016). These features make it reasonable to investigate how hierarchical and modular network structures might influence dynamics near the critical point.

How would activity propagate in a hierarchical, modular network? To get some intuition about this, consider the network shown in figure 7.8D, which consists at its smallest level of clusters, each containing five nodes. Notice that there is a nested structure with clusters of clusters at higher levels. Because of the dense connectivity *within* a cluster, activation of two neurons, converging on a third, can quickly lead to activation of three and then five. This makes it easier for the smallest clusters to become fully active; the mutual connectivity behaves like an amplification mechanism when activity levels are low (Nematzadeh et al. 2014). In contrast, the relatively sparse connectivity *between* the clusters of clusters limits spreading at larger scales, behaving like a damping mechanism when activity levels are high. The net result is that activity is boosted at low control parameter values and reduced at high values, thus stretching the range over which the network will behave in a nearly critical manner.

For contrast, let us compare this to the case of a randomly connected branching network (figure 7.8B), where there is a sharp transition between phases at the critical point governed by the control parameter σ. When activity levels are low, there is no amplification, so the

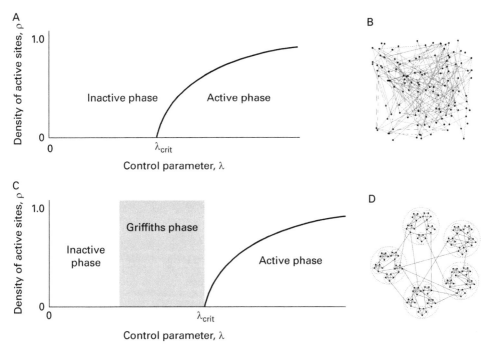

Figure 7.8
Griffiths phase caused by network structure. *A*, Conventional phase diagram for a generic critical model. Control parameter is given by λ and is not necessarily the branching ratio; it could be connection strengths. Transition from inactive to active phase occurs at λ$_{crit}$. *B*, A randomly connected network. *C*, Phase diagram with Griffiths phase given by gray box. There, nearly power-law distributions will occur for a range of control parameter values. The width of the Griffiths phase is exaggerated for clarity. *D*, Hierarchical modular network structure that stretches the apparent critical zone. Network structure in D is adapted from Kaiser, Goerner, and Hilgetag (2007).

branching ratio must reach 1 for propagation to begin. When the branching ratio is high, activity spreads quickly through the entire network, unhindered by barriers between modules. Thus, any branching ratio over 1 will produce supercritical network activation. This makes the phase transition region extremely narrow around the critical point of σ = 1. The same would be true in a more standard integrate and fire neuron model where the control parameter λ might, for example, be the inverse of the firing threshold.

Reports of stretching this apparently critical region have come from both simplified activation models (Kaiser, Goerner, and Hilgetag 2007; Wang and Zhou 2012) as well as more biologically realistic models (Rubinov et al. 2011), so it seems to be a robust phenomenon. There is also experimental evidence that dissociated cultures with different network structures can show a range of critical exponents (Yaghoubi et al. 2018). A similar type of broadening of the phase transition region was discovered by Griffiths when he studied a randomized version of the Ising model (Griffiths 1969). Much later, this work by Griffiths was generalized to complex networks (Muñoz et al. 2010) and then applied more formally to neuroscience by Miguel Muñoz and colleagues (Moretti and Muñoz 2013; Hilgetag and Hütt 2014), firmly establishing the idea that hierarchical modular network structure can broaden the apparent critical zone. In addition, Friedman and Landsberg observed that hierarchical modular networks could even produce avalanche shape collapse and a satisfactory exponent relation when the spreading dynamics did not have a branching ratio of 1 (Friedman and Landsberg 2013). Clearly, the network structure can very strongly impact dynamics near the critical point.

We can now try to map the idea of Griffiths phases onto effective exponents scattered along the line. Because hierarchical modular network structure can allow both subcritical and supercritical networks to appear as if they are nearly critical, we might expect to see a range of exponents—this is consistent with the experimental results, where a clear universality class is difficult to identify. But for the exponents to move along this line, there would need to be changes in network structure. While short-term synaptic plasticity could produce functional network changes on the scale of seconds, long-term synaptic plasticity would need to occur for there to be changes on the scale of minutes or hours. Experiments testing this have not yet been done, but could be performed soon, as they are well within the technical capabilities of many labs.

Most broadly, the Griffiths phase is yet another idea to explain how cortical networks could reap the benefits of criticality without having to be finely tuned to the critical point. It thus provides a margin of error that could be very helpful when biology fails to be precise. Also, a Griffiths phase is not mutually exclusive of quasicriticality or the subsampling idea. It is conceivable that several of the proposed mechanisms are operating together to help optimize information processing in cortical networks, even as they are constantly driven by external stimuli and undergoing changes in synaptic strengths.

Chapter Summary

Experiments outlined in chapter 6 showed that living neural networks are constantly perturbed away from the critical point and moving back toward it through homeostasis. In this chapter, we examined how these networks operate in the close vicinity of the critical point. The fact that critical exponents continually move along the exponent line gives us a clue to what might be happening.

We first described the idea of quasicriticality. There, external drive causes a network to depart from the critical point, moving in the region along a line where susceptibility and mutual information are still optimized. In this idea, the concept of universality as an ideal limit is still preserved. Data from recent experiments are consistent with quasicriticality's predictions. Second, we looked at the idea that these networks are slightly subcritical, maximizing information processing without venturing into an unstable regime. Here too, experimental evidence is consistent with this idea, but we noted that quasicriticality makes some different predictions that have yet to be tested. Third, we explored how movement along the exponent line could be explained by subsampling, and we described an experiment that could put this idea to the test. Fourth, we discussed the Griffiths phase, where the apparent critical regime could be stretched by network structures that are modular and hierarchical. All these ideas are currently plausible, and they are not necessarily mutually exclusive. Future research will be needed to distinguish their predictions more carefully before a more definitive picture can emerge.

Exercises for this chapter can be found through a link given in the appendix.

8

Cortex

Good mathematicians see analogies. Great mathematicians see analogies between analogies.
—attributed to Stefan Banach by Stanislaw Ulam

Breadth of training predicts breadth of transfer. That is, the more contexts in which something is learned, the more the learner creates abstract models, and the less they rely on any particular example. Learners become better at applying their knowledge to a situation they've never seen before, which is the essence of creativity.
—David Epstein, *Range: Why Generalists Triumph in a Specialized World*

Compared to other species, humans are physically unimpressive. We're much slower than cheetahs, squirrels, and dogs; we don't have fangs or venom; most of us can't even do one chin-up; and we're really only comfortable when the temperature is between 22°C and 24°C. Our offspring are helpless and dependent for a decade or more, while newborn foals get up and walk two hours after being born. Apparently, all of this can be overcome just by having the right kind of brain—this is what makes us apex predators. Who needs fangs or quickness when you can throw a sharp spear and work in a team?

In this chapter, I want to explore how the expansion of the cortex increased our intelligence.[1] I will argue that this expansion, by itself, would not have improved intelligence—it required the newly added regions of cortex to operate near the critical point. This is admittedly a speculative idea, but we are in the part of the book where we are most focused on potential future directions. I am inspired by notable neuroscientists of the past who were willing to stick their necks out and guess what would come next. Santiago Ramón y Cajal famously drew arrows in his diagram of the hippocampus to show how he thought information flowed through it. He did this at a time when Golgi and others still thought synapses didn't exist and that all neurons formed a continuous reticulum (Hellman 2001). Much later, after synapses were confirmed, Donald Hebb wrote about how their strengths might be increased by coactivity, as he described the learning rule that would later bear his name. He also went on to describe how things he called "cell assemblies" could store memories (Hebb 2005). None of these ideas would have likely passed peer review or the scrutiny of a funding panel, yet today there is overwhelming evidence for their correctness (Kelso, Ganong, and Brown 1986; Yeckel and Berger 1990; Harris et al. 2003). As there is a thin line between being prophetic and sounding like an idiot, I must tread carefully. While I am no Cajal or Hebb, I can be inspired by them and try to emulate their thoughtfulness and courage. Experiments will eventually tell us if the ideas in this chapter were on the right track or not.

In what follows, we will first describe the expansion of the neocortex and then some of its circuit-level characteristics. From there, we will discuss the important role of association cortices in forming deep hierarchies for processing more abstract information. I will argue that cortical layers 2 and 3 played a pivotal role in allowing this to happen and that they in particular should operate near the critical point. We will close with a look at some data that are consistent with these ideas, as well as a few predictions for future experiments.

The Expansion of Cortical Area

I have always wanted to know what makes some brains more intelligent than others. For example, why is Homo sapiens able to do quantum mechanics while presumably our ancestor Australopithecus could not? The most obvious difference between us and them, and the first answer that springs to mind, is that we have much larger brains (Hofman 2014). Australopithecus had a cranial volume of about 500 cm^3, slightly larger than our closest living relative, the chimp, at 400 cm^3. In contrast, we have a volume about three times that, 1500 cm^3 (figure 8.1A) (Schoenemann 2006). If we look specifically at the size of the neocortex,[2] the brain's crinkled outer layer of neurons, we see that a human has almost four times the cortical surface area of a chimpanzee (figure 8.1B). Even though we may share 99 percent of our genes with chimps (Mikkelsen et al. 2005), there is not merely a 1 percent difference between us here; we have nearly 300 percent more cortex than they do.

While we have dramatically more cortical area, the differences in cortical structure between us seem to be modest. In fact, a cross-section of the cortex reveals it to have six layers (figure 8.2) that are remarkably similar across mammals from rats to humans (O'Reilly and Munakata 2000; Hutsler, Lee, and Porter 2005). To understand these similarities, it will be helpful to briefly overview the cellular composition of neocortex and its connectivity.

In terms of cell populations, most studies have recognized three main populations of neurons (Douglas, Martin, and Whitteridge 1989). Layers 2 and 3 are filled with small excitatory pyramidal neurons, and layer 4 has many star-shaped (or stellate) inhibitory neurons, mixed with excitatory neurons of various shapes. Layers 5 and 6 have excitatory pyramidal neurons, some of whose dendrites extend to the upper layers (figure 8.2B) (Lefort et al. 2009; Harris and Mrsic-Flogel 2013; Harris and Shepherd 2015). Keep in mind that this is a simplified picture; entire books are written about subtle features of cortical structure (White 1989; Braitenberg and Schüz 1998). Nonetheless, the similarities across species are real and consistently noted (Hutsler, Lee, and Porter 2005).

Regarding connectivity, inputs from the thalamus as well as other cortical areas arrive in layer 4. From there, layer 4 neurons send their strongest projections to layers 2 and 3. The layer 2 and 3 pyramidal neurons send projections out to other cortical areas. They also send projections to the pyramidal neurons of layers 5 and 6 (Lefort et al. 2009; Markram et al. 2015). These deep layer neurons send connections to subcortical structures like the thalamus or striatum (figure 8.2C). Again, this is the simplified picture. To

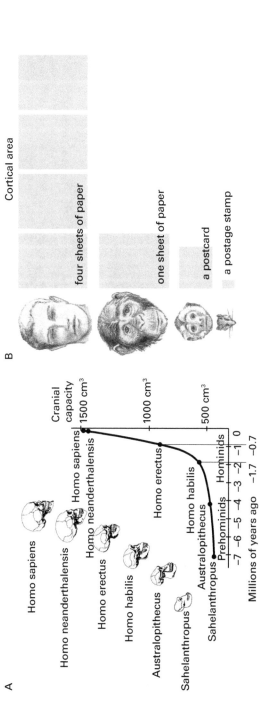

A

Homo sapiens

Homo neanderthalensis

Homo erectus

Homo habilis

Australopithecus

Sahelanthropus

Cranial capacity

1500 cm³

1000 cm³

500 cm³

Homo sapiens
Homo neanderthalensis

Homo erectus

Homo habilis

Australopithecus

Hominids

Prehominids

Sahelanthropus

Millions of years ago

−7 −6 −5 −4 −3 −2 −1 0
−1.7 −0.7

B

Cortical area

four sheets of paper

one sheet of paper

a postcard

a postage stamp

Figure 8.1

Expansion in cranial capacity and cortical area. *A*, Cranial capacity, or the volume of the entire brain, is much larger in humans and Neanderthals than in our other hominid ancestors. *B*, The expansion of cranial capacity was accompanied by an expansion in cortical surface area. When the convoluted cortex is unfolded, it shows that human cortical area is about four sheets of paper. A chimpanzee has about one sheet of paper; a capuchin monkey has about a postcard; a rat has about a postage stamp. Panel A is adapted from Bretas, Yamazaki, and Iriki (2020); panel B is adapted from Calvin (1994).

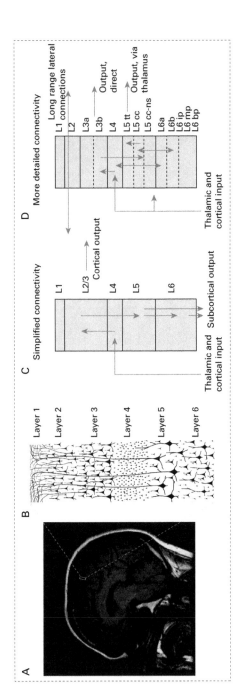

Figure 8.2

A six-layered structure is similar throughout cortex. *A*, Magnetic resonance image of the author's skull and brain; we will zoom in to see a cross-section through the cortex near the small rectangle. *B*, Schematic of cortical neurons arranged in six layers that would appear in the rectangle. Although there are some important exceptions, most regions of neocortex would show a roughly similar structure. *C*, Simplified circuit diagram; this general pattern is followed across mammalian species (Garey 1999). Inputs arrive in layer 4, and cortico-cortical connections are sent from layers 2 and 3. *D*, A more detailed circuit diagram, showing complexity. Panel B is adapted from Barrett (2020); panels C and D adapted from Hawkins, Ahmad, and Cui (2017).

give you some idea of detailed elaborations on it, look at figure 8.2D. For our purposes, the main point now is that layers 2 and 3 are primarily responsible for sending connections to other cortical areas.

Associations of Associations

If the layered structure of the neocortex is relatively similar between us and chimpanzees, then the main difference between our cortices would seem to be in terms of area. This prompts the question: Why would enlarged cortical area be so important for amplifying intelligence? To get an idea about this, let us compare how cortical area is allocated in rats, cats, monkeys, and humans (figure 8.3). It has long been known that sensory functions like vision, touch and hearing are dependent on specific cortical regions; these regions are known as sensory cortex. Likewise, motor cortex is necessary for controlling movement of muscles. In the relatively small brain of the rat, most of the cortical territory is consumed with sensory and motor functions. This leaves proportionately little cortical area left over for associations between sensory modalities, or associations between senses and motor actions. For example, sensory associations between vision and audition could include the observation that dogs make barking sounds or that cats meow, while monkeys howl. An example of relating touch with taste would be that soft bananas taste better than hard ones. With increased cortical area, more associations can be formed, so more relationships in the world can be noticed. This clearly could help with survival.

In addition to *more* associations, though, increased cortical area gives space for *higher-order* associations to form. For example, the previous first-order associations of monkeys with howling sounds and softness with bananas could themselves be combined to create this second-order association: those howling sounds might mean that monkeys are coming for the ripe bananas before me. I should eat them now or pick them up and run. With enough cortical territory, third-order and fourth-order associations can form, leading to a hierarchy among cortical regions (figure 8.4).

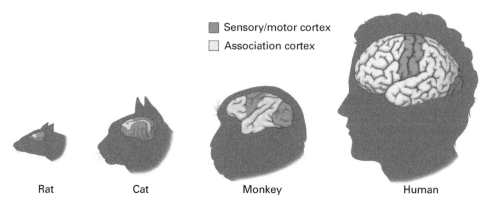

☐ Sensory/motor cortex
☐ Association cortex

| Rat | Cat | Monkey | Human |

Figure 8.3
Larger brains have more association cortex. Sensory and motor cortices are shown in dark gray, while association cortices are in light gray. In the rat, most cortical area is occupied by essential sensory and motor functions. In humans, sensory and motor areas still occupy substantial territory, but the larger brain gives proportionately more area to association cortices. Adapted from https://slideplayer.com/slide/14149409.

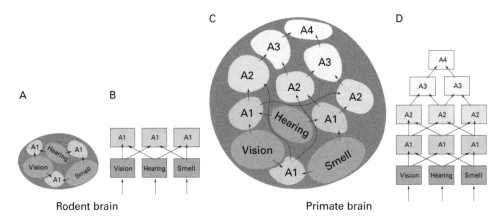

Figure 8.4
More association cortex allows higher-order associations. *A*, Schematic of a rodent brain, with little cortical territory left for first-order (A1) associations between sensory cortices. *B*, With the first-order associations, it forms a network of two stages. *C*, Schematic of primate brain with ample area for association cortices. Here, second- (A2), third- (A3), and fourth-order associations (A4) can form. *D*, A hierarchical network with five stages is formed.

In primate brains, these types of hierarchies are indeed found. The visual system is the most thoroughly studied example of this: primary visual cortex V1 detects edges and lines, V2 detects patterns of lines, V4 detects shapes of intermediate complexity, and inferior temporal cortex, IT, can detect complex objects including faces (Felleman and Van Essen 1991) (figure 8.5A). Each higher cortical region represents combinations of features from regions below it, permitting increasingly complex representations to be built up. Another hierarchy exists for the somatomotor system (figure 8.5B). Just as Banach noted that great mathematicians see analogies of analogies, more intelligent animals make associations of associations. And this can only happen if there is enough cortical area to allow it.

Although it might seem intuitively obvious that higher-order associations would lead to greater intelligence, we can further support this claim by looking to recent results in the field of deep learning. There, artificial neural networks loosely inspired by the brain have shown tremendous gains in processing power compared to their earliest days (Rosenblatt 1958). Starting around 2011–2012, artificial neural networks began to outperform humans on pattern recognition tasks (Ciresan et al. 2012). There soon followed major improvements in automatic speech recognition (Hannun et al. 2014), superhuman performance in the game of Go (Silver et al. 2016) and a breakthrough in the protein-folding problem (Heaven 2020). This revolution was primarily driven by the increased number of processing stages, or layers,[3] in these networks (Krizhevsky, Sutskever, and Hinton 2012). More stages, or depth, were only possible after substantial improvements in computer hardware like faster central processing units and parallel processing in graphics cards.

Beyond the technological fact that more processing stages just seem to work better, there is also a theoretical justification. To explain this, we will first describe more broadly what an artificial neural network does. Briefly, it takes some set of inputs and maps them onto a set of outputs. For example, the inputs could be a set of pictures and the outputs could be "yes" or "no" answers to whether there was a dog in the picture; the inputs could be a set of noisy recordings of people's voices and the outputs could be de-noised recordings.

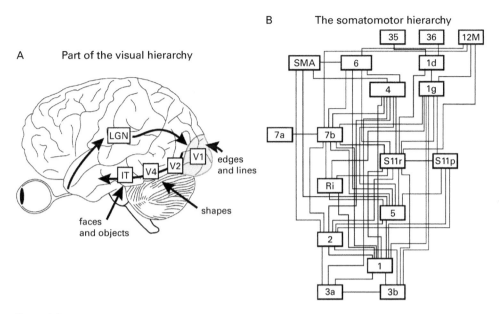

Figure 8.5
Higher-order association cortices are arranged in hierarchies. *A*, Cortical regions in ventral stream of human visual system represent progressively more complex items (adapted from Herzog and Clarke [2014]). *B*, Somatosensory and motor cortical regions form a hierarchy in macaque brain (adapted from Shipp [2005]).

More generally, such an input-output mapping can be represented by a mathematical function, like output = f(input), where the function f could be, for example, some polynomial. Under reasonable assumptions, a shallow neural network with only one processing layer[4] is sufficient to approximate any function (Barron 1994), but the single layer that does this may require enormous numbers of neurons (Lin, Tegmark, and Rolnick 2017). Several authors have shown that deeper networks with more layers allow functions to be represented with fewer neurons (Bianchini and Scarselli 2014; Rolnick and Tegmark 2017). Not only does having fewer neurons conserve resources, it also reduces the number of connections to adjust, meaning that training requires fewer repetitions. In other words, deeper networks allow brains to be more powerful within a given volume, and they make learning faster. The hierarchy of cortical processing stages has been compared to the multiple layers of deep artificial neural networks (figure 8.6) (Yamins et al. 2014).

It may help to quickly summarize the argument so far. Higher intelligence seems to be driven by more cortical area. This increased area allows enough association cortex to form hierarchies, and these hierarchies can perform more complex computations. Having more association cortex is like having more layers in an artificial neural network, making it deeper. Amplifying intelligence thus crucially depends on having more cortical regions, and having those regions connect to each other in a hierarchy with depth. I now want to explore how increased connectivity between cortical regions came about. This will lead us to consider the crucial role that being near the critical point plays in amplifying intelligence.

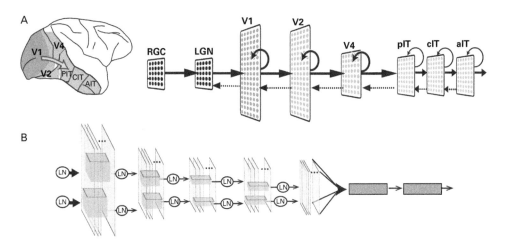

Figure 8.6
The hierarchy of visual regions is analogous to the many processing stages of a deep neural network. *A*, At left, macaque brain with ventral visual stream in gray. At right, each stage in the stream is represented as a population of neurons arranged in a sheet: RGC, retinal ganglion cells; LGN, lateral geniculate nucleus of the thalamus; V1, primary visual cortex; V2, secondary visual cortex; V4, visual area V4; pIT, posterior inferior temporal cortex; cIT, central; aIT, anterior. Straight solid arrows schematically represent feed-forward connections. Curved solid arrows are recurrent connections. Dashed arrows are feedback connections. *B*, Deep neural network model of the visual stream, constructed to perform object recognition and minimize wiring cost. LN represents linear-nonlinear units that mimic operations performed in populations of neurons with nonlinear thresholds. After training, response properties of artificial neurons at each stage of the network were very similar to response properties found in analogous stages of the visual stream (Yamins et al. 2014; Yamins and DiCarlo 2016). Panels A and B adapted from Lee et al. (2020).

The Special Role of Layers 2 and 3

Let us return to the comparison between cortical area in humans and chimpanzees. If human cortex has about four times the area of chimp cortex, what does this imply about the connectivity that must exist between cortical regions? To answer this, let us do a thought experiment with a small number of cortical regions, say one for each 100 cm³ of cortex. Since the chimp has about 400 cm³, it would have four regions. Since the human has nearly 1600 cm³ if we round up, let us say it has 16 regions. Now, if we want each region to connect to every other region, and make recurrent connections with itself, we can calculate how many connections chimps and humans should have. For the chimp, each region needs to connect to its three neighbors and itself, so that would be four connections per region. In total, this would be four connections for four regions, or $4^2 = 16$ connections. For the human, each region would need to connect to its 15 neighbors and itself, making 16 connections per region. For 16 regions, this would be $16^2 = 256$ connections. This means that the number of connections goes up as the square of the area. An increase in area by a factor of four means an increase in connectivity by a factor of 16 ($C_{human}/C_{chimp} = 256/16 = 16$).

This steep increase in connectivity had to be accommodated somehow. Interestingly, there are modifications in the standard six-layered cortical structure as we move to species with more cortical area. From our earlier description, we know connections to other cortical regions are primarily sent by layers 2 and 3 (figure 8.2C). As cortical area increases, we would expect these layers to be sending more connections than before. If we look at

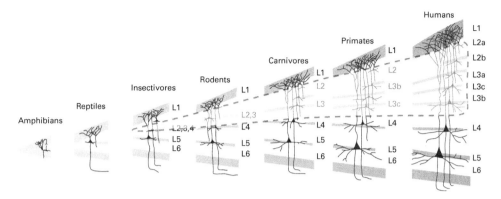

Figure 8.7
Increased cortical area is accompanied by thicker layers 2 and 3. Schematic cross-section of cortex for amphibians to humans. Note dashed triangular area showing how layers 2 and 3 are continually elaborated by the addition of sublayers like L2a, L2b, L3a, and so on. In contrast, layers 4, 5, and 6 remain relatively constant in thickness. Adapted from Hill and Walsh (2005).

cortical cross-sections for various species, we do find that layers 2 and 3 become thicker as overall brain size is increased (Hill and Walsh 2005; Hutsler, Lee, and Porter 2005). The other layers remain relatively unchanged in thickness (figure 8.7).

From this perspective, the increased connections from layers 2 and 3 were essential for the development of deeper cortical hierarchies. If this is true, then learning in these hierarchies would depend on synaptic plasticity in these layers. Consistent with this, the density of receptors implicated in synaptic plasticity (NMDA, AMPA mGluR2/3 receptors) is greatest in layers 2 and 3 across a wide variety of cortical regions (Palomero-Gallagher and Zilles 2019). These layers are known to show high levels of plasticity, at least in sensory cortex (Feldman and Brecht 2005; Diamond, Huang, and Ebner 1994). Layers 2 and 3 therefore are likely to play a key role in capturing the associations experienced in the environment and wiring up the hierarchy to reflect the statistics of the world. Interestingly, the relative thickness of Layers 2 and 3 changes with different types of cortex: they are thinner in sensory and motor cortices and thickest in association cortices (Shepherd 2004). This agrees with the hypothesis that the connections from association regions of cortex are responsible for learning in deeper hierarchies. Now let us move on to consider how the task of forming associations is related to multifunctionality, and how multifunctionality may be related to the critical point.

Multifunctionality and the Critical Point

In addition to this capacity for plasticity, an association-forming brain circuit would also need to be highly unspecified, a relatively blank slate. This is because most associations cannot be predicted beforehand—they must be experienced to be learned (Wise and Murray 2000). For example, if you grow up in the United States, you drive on the right side of the road; things are different in England. Most word-object associations are likewise arbitrary. A capuchin monkey can learn that picking up the blue block leads to a cashew reward while the yellow disk brings a raisin. A circuit capable of learning these arbitrary associations must be highly flexible and therefore minimally constrained.

This pluripotent role for layers 2 and 3 in the cortex is also something that is seen in neocortex more generally. Recall that the six-layered structure of the cortex is used for very diverse tasks, ranging from processing visual and auditory sensory information, to commanding motor actions, to planning complex sequential tasks. As we noted before, there is also evidence that cortical function can be altered by the inputs it receives: Sharma and colleagues showed that when visual inputs were rerouted to auditory cortex, it caused auditory cortex to develop receptive fields reminiscent of those seen in visual cortex (Sharma, Angelucci, and Sur 2000). The layered cortical circuit is highly versatile.

Further evidence for the multifunctionality of the cortex can be based on its relative recency in evolution and development. Evolutionarily, the neocortex is the last structure to be elaborated and because of that, it is the least specified. From the standpoint of development, neurons forming the outer layers of the neocortex, layers 2 and 3, are among the last to migrate outward along radial glia and settle in their positions (Rakic 2009). The connections formed by these neurons remain plastic throughout the lifetime of the organism, ready to learn those seemingly arbitrary associations. This trend of multifunctionality and delayed development is also clearly seen in the hierarchy of brain regions. As scientifically literate parents often tell their teenagers, connections with the prefrontal cortex, the highest regions in the hierarchy, do not mature in humans until their early twenties. In contrast, primary visual cortex, the lowest region in the hierarchy, goes through an early critical period so that vision can be established as soon as possible (Shaw et al. 2008). These trends suggest those circuits that remain plastic longest and mature latest are the ones that are the most multifunctional. And it is these multifunctional areas that are most responsible for the associations of associations.

Now, the multifunctionality of the cortex brings with it certain requirements that lead toward the critical point. Consider the association-forming circuits in layers 2 and 3 of the cortex. Certainly, a desirable feature for such circuits would be a large memory capacity—the more associations they could store, the better. Recall that the branching model shows a peak in the number of stable spatiotemporal patterns, like those linked to memories, near the critical point (figure 4.11). This alone would suggest that layer 2 and 3 cortical circuits would benefit from operating near the critical point. But in addition to storing memories, multifunctional circuits would also be called upon to transmit information through the hierarchy with minimal loss, to be sensitive to sight changes in inputs, to perform computations optimally and to be stable yet controllable. In short, because these circuits would need to be ready for anything, they would need to be good at all things. They would have to be highly flexible computational devices. All these possible roles would need to be optimized in them simultaneously. From our previous discussions in chapter 4, we know that this can be satisfied when the circuit operates near the critical point (figure 8.8).

This would not necessarily be the case in circuits that were specialized for a particular task. For example, consider the rhythmic pre-Bötzinger complex in the brainstem that regulates breathing (Smith et al. 1991). The function of this circuit is to reliably generate signals that have roughly the same period, modulated by some physiological demand like lack of oxygen (Feldman and Del Negro 2006). A similarly specialized circuit is formed by the delay lines in the nucleus laminaris of the barn owl (Carr and Konishi 1990; Carr and Boudreau 1993). These are thought to allow them to detect microsecond arrival time differences of sounds in their ears (Wagner et al. 2005); this is undoubtedly handy for

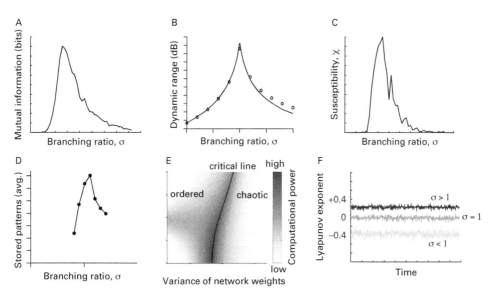

Figure 8.8
Functions optimized simultaneously near the critical point. Figures shown previously are here presented together in schematic form. *A*, Mutual information peaks near a branching ratio of 1. *B*, Dynamic range also peaks near the critical point. *C*, Susceptibility shows a similar peak. *D*, Number of stored patterns, a proxy for memory, peaks near a branching ratio of 1. *E*, Computational power peaks along the critical line in a model that can be tuned by the variance of the connection weights. *F*, A Boolean version of the Lyapunov exponent (Derrida and Pomeau 1986) is near zero when the branching ratio is 1, indicating neutral (stable yet controllable) dynamics. A positive Lyapunov exponent is associated with chaotic dynamics and a negative one with attractive dynamics. See the appendix for a more detailed explanation. B is adapted from Kinouchi and Copelli (2006); D and F are adapted from Haldeman and Beggs (2005); E is adapted from Bertschinger and Natschlager (2004).

locating mice at night. We have no reason to suspect, though, that either of these circuits would be good at learning arbitrary associations, or that they would maximize information storage capacity. Their job is to do one thing well, often at the expense of all other things. The more specialized a circuit becomes, the less it would need to optimize multiple information processing functions simultaneously. For this reason, we would not expect these circuits to operate near the critical point.

This intuition is corroborated by modeling studies. Cramer and colleagues used a neuromorphic chip to simulate a neural network that could be tuned around the critical point (Cramer et al. 2020). They challenged it with computational tasks of varying complexity, from calculating the sum of inputs delivered over five time steps (simple) to calculating parity[5] over twenty-five time steps (complex). In each of these calculations, the network had to store the inputs in memory and then process the information. The expectation was that critical networks would perform better when long duration memory was required, as information decays most slowly in systems near the critical point. For complex tasks, being near the critical point indeed maximized performance. In contrast, subcritical networks performed best on simple tasks. The authors speculated that holding information in memory was detrimental for simple tasks because it consumed network resources. Studies using digital evolution came to similar conclusions and found that the subcritical regime was best for simple tasks while complex tasks required proximity to criticality (Villegas et al. 2016; Prosi et al. 2021).

If this idea about specialization is correct, then we would expect the most recent corti-cal layers and regions to be the most flexible and multipurpose, capable of forming unan-ticipated associations and performing complex tasks. Those cortical regions and layers that crystallize earlier would be less flexible and more specialized and not necessarily as close to the critical point. The prediction of this hypothesis is that signatures of being near the critical point should be strongest in the parts of the cortex that are most recently added. Let us now see if the data agree with this general pattern.

Nearly Critical in Layers 2 and 3, but Not in Layer 5

One of the earliest observations about neuronal avalanches was that they occurred in lay-ers 2 and 3 of acute cortical slices (Beggs and Plenz 2003). This result was made clearer by Craig Stewart and Dietmar Plenz, who noted that the LFP signals caused by neuronal avalanches showed the highest density in layers 2 and 3 (Stewart and Plenz 2006). Further work by Plenz and colleagues showed that neuronal avalanches occurred in layers 2 through 4 of mouse primary auditory cortex in vivo in response to sounds (Bowen et al. 2019); they did not examine layer 5 neurons, though.

Slightly more recent work by Ma, Peters, and colleagues used calcium imaging to rec-ord neuronal avalanches in vivo from about 200 neurons at a time in primary motor cor-tex, M1, of the mouse (Peters et al. 2017; Ma et al. 2020). Using this technique, they were able to simultaneously record activity from pyramidal neurons in layers 2/3 as well as the dendrites of layer 5 pyramidal neurons that extended into layers 2/3. Previous controls had established a firm link between the calcium signal in these apical dendrites and the spiking of their cell bodies (Peters et al. 2017). They found that the cascades of activity in the layer 2/3 neurons followed power-law distributions, and that the exponents of these distributions obeyed the exponent relation. In addition, the avalanches had a characteris-tic inverted parabola shape that was similar across different scales. The exponent used to collapse avalanche shapes was derived from the power law produced by the data, where $\gamma \approx 1.25$. All these interlocking signatures indicated that the layer 2/3 neurons were oper-ating near the critical point (figure 8.9A, B).

The neurons they recorded from layer 5, however, did not show these signatures. While there was a power law relating average avalanche size to avalanche duration, it did not satisfy the exponent relation (figure 8.9C). Moreover, the avalanches produced did not have an inverted parabola shape, and they did not have good collapse (figure 8.9D). The data showed that the layer 5 neurons were not operating near the critical point, a finding consistent with the earlier work done on acute cortical slices.

Why would layer 5 neurons differ from layer 2/3 neurons in this respect? As mentioned above, layer 2/3 neurons are evolutionarily and developmentally more recent, and their functions may be more unspecified than layer 5 neurons. In this experiment, layer 5 cortico-spinal cells were selected to express the calcium indicator (Ma et al. 2020; Peters et al. 2017), so other types of layer 5 neurons were not imaged. These neurons have the tallest dendrites and receive connections from every cortical layer, outputting bursts of spikes to distant subcortical targets (Harris and Mrsic-Flogel 2013). It may be that the main task of these neurons, encoding output for motor control, is more specialized than forming gen-eral associations, and that optimizing this occurs at the expense of other tasks. Such reduced

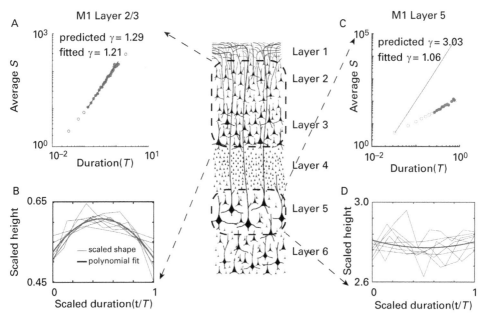

Figure 8.9
Layer 2/3 neurons operate close to the critical point, but layer 5 neurons do not. *A*, Average avalanche size for a given duration plotted against avalanche duration for layer 2/3 neurons from primary motor cortex, M1, of the mouse. The slope from the exponent relation (predicted γ) plotted as a straight line with data points (fitted γ) shows close agreement. *B*, Avalanche shapes are inverted parabolas and collapse well for activity from layer 2/3 neurons. *C*, Layer 5 neurons show mismatch between predicted and fitted γ. *D*, Avalanche shapes are flatter and show poor collapse. Data panels adapted from Ma et al. (2020); cortical column adapted from Barrett (2020).

multifunctionality may make operating near the critical point unnecessary for them. Clearly more research will be needed to understand this issue, but it is intriguing that even within the cortex, there appear to be differences in proximity to the critical point.

Staying Nearly Critical While Learning

Another interesting feature of this study was that the mice were imaged while they learned to press a lever for a water reward (Ma et al. 2020). This allowed the investigators to assess proximity to the critical point over two weeks during training. In this period, they found that the activity of layer 2/3 neurons eventually became more consistent with movements, generating reproducible spatiotemporal patterns (Peters, Chen, and Komiyama 2014). Layer 5 neurons, however, showed a different trajectory—there, the activity associated with dissimilar movements became more decorrelated (Peters et al. 2017). It is as though the job of the layer 2/3 neurons was to learn the sequence required for lever press, while the job of the layer 5 neurons was to clearly separate output signals that might have been overlapping. These results point to very different functions of the cortical layers, and may help to make sense of why one, but not the other, operates near the critical point.

When the neuronal avalanches produced by layer 2/3 neurons were analyzed over multiple training sessions, they found that the exponents changed considerably. Interestingly, while there was much variability, these exponents continued to adhere to the exponent

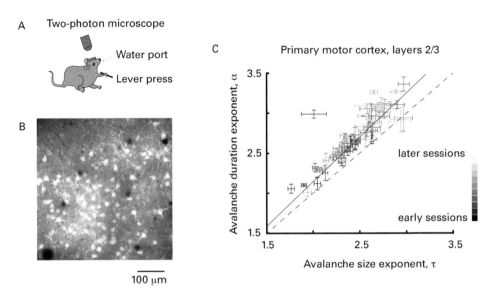

A Two-photon microscope

Water port

Lever press

B

100 μm

C Primary motor cortex, layers 2/3

later sessions

early sessions

Avalanche duration exponent, α

Avalanche size exponent, τ

Figure 8.10
Layer 2/3 neurons remain close to critical during learning. *A*, Mice were trained to push a lever for water reward and were monitored for 14 days. *B*, Field of view for head-fixed two-photon microscope, showing neurons in primary motor cortex, caudal forelimb area. About 200 neurons were imaged per mouse ($N = 7$ mice), and the same neurons could be tracked over the course of two weeks. *C*, Exponents from avalanches produced in cortical layers 2/3 changed over time. Early sessions are darkest; later sessions lighter. Data for all 7 mice over 14 days shown. Lines drawn for reference are dashed: $\gamma = 1$ and solid: $\gamma = 1.28$. All panels adapted from Ma et al. (2020); data for these analyses was originally produced by Peters et al. (2017).

relation (figure 8.10). These findings are consistent with those of Shew et al. (2015) in turtle cortex and Fontenele et al. (2019) in several species, who reported that many independently measured exponents still obeyed the exponent relation. Here, though, the exponent relation was followed in the same animals as primary motor cortex underwent substantial synaptic reorganization. Despite these large changes, the layer 2/3 neurons continued to operate near the critical point. Layer 5 neurons did not show signatures of criticality, even after learning.

The data from cortical layers thus support the hypothesis that multifunctionality and recency are correlated with operating near the critical point. Are there any data to indicate that cortical regions higher in the hierarchy are closer to the critical point than those lower in the hierarchy? Let us now turn to a study on correlation times to address this question.

Timescales throughout the Hierarchy

Back in chapter 1, where we first discussed the implications of operating near the critical point, we saw that information decayed most rapidly in systems that were subcritical and supercritical. Systems near the critical point had information die out with longer time constants (figure 1.5). Indeed, one of the first experimental results to suggest that the brain may be operating near the critical point came from Linkenkaer-Hansen and colleagues, who showed that long range temporal correlations of the MEG signal had a power law decay in human subjects (Linkenkaer-Hansen et al. 2001) (figure 1.10). From this, we would expect that cortical regions closer to the critical point should show longer decays of information over time.

One way to measure this type of decay at the neuronal level is by examining autocorrelations in spike trains. An autocorrelation plot shows the correlation between one spike and another, from the same neuron, across different delays. If the autocorrelation drops rapidly, say in 3 ms, then a spike at one time is not very predictive of later spikes; if it drops gradually, say in 300 ms, then one spike is likely to be followed by other spikes for about a third of a second, which is a relatively long time for a neuron. Figure 8.11B shows three example autocorrelation curves. These curves can typically be approximated by a decaying exponential function with a time constant, τ, not to be confused with the critical exponents we discussed earlier. The value of τ gives the time at which the autocorrelation curve decays to about 37 percent of its maximal value, and this is often used as a handy way to describe timescales in spike trains.

To assess the intrinsic processing times of different cortical regions, Murray and colleagues averaged autocorrelation time constants for many neurons from each region (Murray et al. 2014). They did this by pooling results from seven different primate labs to ensure that the data did not merely reflect local conditions in the labs. The overall trend was that cortical regions higher in the hierarchy had significantly longer time constants than those lower in the hierarchy (figure 8.11C). These cortical regions also have more abstract representations (Badre and D'Esposito 2009) and larger receptive fields (Lennie

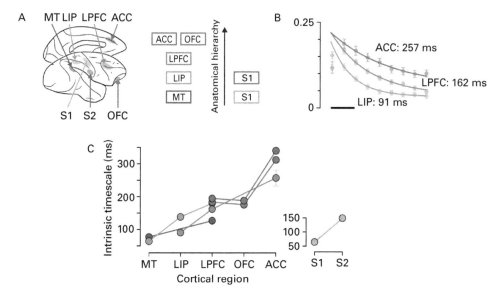

Figure 8.11
A hierarchy of cortical timescales. *A*, Cortical regions that were compared: MT, medial-temporal cortex; LIP lateral intraparietal cortex; LPFC, lateral prefrontal cortex; OFC, orbitofrontal cortex; ACC, anterior cingulate cortex; S1, primary somatosensory cortex; S2, secondary somatosensory cortex. Based on patterns of connections, they form an anatomical hierarchy (Felleman and Van Essen 1991; Barbas and Rempel-Clower 1997). *B*, Representative autocorrelations averaged from spike trains from three regions, showing differences in decay over time. Each data set can be fit by an exponential curve with the time constant shown, giving the intrinsic timescale. Error bars indicate standard error of the mean. *C*, When data were pooled from six research groups, the regions highest in the hierarchy had the longest intrinsic timescales ($P < 10^{-5}$, $r_s = 0.89$, Spearman's rank correlation). Adapted from Murray et al. (2014).

1998) than regions below them, in keeping with our discussion that they form associations of associations. This result matches our expectation that the most multifunctional cortical regions are the closest to the critical point. It will be illuminating to see if measures of the branching ratio in these regions show a parallel trend, with the highest areas being the closest to the critical value of $\sigma = 1$.

These findings are part of a larger literature that has highlighted the multiple time-scales of processing in the brain (He et al. 2010; Honey et al. 2012; Chen, Hasson, and Honey 2015). Models have shown that this hierarchy of timescales can emerge among cortical regions as a result of their local recurrent connections and their interconnections between each other (Chaudhuri et al. 2015). Interestingly, a model with a power law decay curve, rather than an exponential curve, has superior performance in memory tasks that link events across multiple time scales (Chien et al. 2021). Relating long timescale integration to the critical point and cognitive performance should be a fruitful avenue for future research.

Chapter Summary

The mystery behind the origins of higher intelligence is unsolved, but fascinating. In this chapter we outlined a hypothesis for how intelligence was amplified through the expansion of neocortical area. According to this view, the crucial factor was not so much that a larger cortex increased the *amount* of processing, but that it increased its *depth*. More area gave space for associations of association cortices, forming a deeper hierarchy of cortical regions. This enabled the cortex to detect abstract relationships in the world that a shallower hierarchy would have missed. In a similar manner, the power of artificial neural networks dramatically increased when they became deeper with more layers.

But greater depth in the cortical hierarchy would not have been possible unless the regions at higher levels were versatile and capable of learning arbitrary associations. Such regions could not be highly prespecified but would require flexibility for the lifetime of the organism. Thus, higher cortical regions had to be consummate generalists, capable of storing information well, transmitting it optimally, performing computations efficiently, being sensitive to slight changes in inputs, and easily controllable yet not chaotic. These regions had to avoid the tradeoffs that would occur with becoming highly specialized— good at one function at the expense of all the others. While being good at many things is difficult, multiple information processing demands can be simultaneously optimized near the critical point. From this perspective, the requirements of versatility drove higher cortical regions toward criticality.

This view has several predictions about what parts of cortex will operate closest to the critical point. Cortical layers 2 and 3 send projections to other cortical regions, and their increased thickness correlates strongly with increased overall cortical area. They are also known to have high levels of synaptic plasticity. These layers therefore appear likely to mediate arbitrary associations within the hierarchy and function as generalists. Because of this, they are predicted to operate near the critical point. Experimental evidence shows that layers 2 and 3 have signatures of being near the critical point; these signatures are absent in layer 5 neurons. Because cortical regions higher in the hierarchy mediate more abstract relationships, they too are predicted to be closer to the critical point than lower sensory regions. While this awaits further exploration, we do see that autocorrelation times are longer in higher regions, as predicted. This hypothesis is still quite speculative, but experiments in the near future could put it to the test.

According to this hypothesis, expansions in cortical surface area by themselves would not have allowed dramatic increases in intelligence. Rather, the most important thing was that the newly expanded areas would operate close to the critical point. The leading edge of intelligence amplification is thus formed by association cortices operating near the critical point. The trailing edge consists of highly specialized circuits, performing already learned but essential functions. This hypothesis predicts that future amplifications of intelligence, whether in cortical evolution or in artificial neural networks, will require components at the leading edge to be nearly critical.

9

Epilogue

My goal with this book has been to give an introduction that would allow someone new to this field to hear its main claims, to see why it is exciting and growing, and to understand its central concepts. This has been written at an introductory level, and there is far more that could have been said. For those who are intrigued and want to go on, I will now summarize what we know, what we don't know, and where the frontier issues are likely to be.

What We Know

These are the main findings conveyed in this book. While it is probably never the case in science that we confidently know anything for sure, I consider the information below to be solid. This does not mean that some would never question it, but that there is an emerging consensus on these issues, as indicated by the direction of the field.

Operating near the critical point: There is now abundant evidence that the cortex often operates near a critical phase transition point. This comes from species ranging from fish to humans and is demonstrated by data from methods ranging from electrode recordings to MEG. There are multiple signatures of being near the critical point, including a branching ratio very close to 1, evidence of tuning between phases, multiple power laws in the phase transition region that pass rigorous statistical tests, an exponent relation that is approximately satisfied and avalanche shape collapse. It is now standard practice to document most of these signatures, and labs around the world are consistently doing so.

Optimizing information processing: There is increasing evidence that operating near the critical point improves multiple forms of information processing, including information transmission, dynamic range, and susceptibility. While this early work started out in vitro, later studies have confirmed that operating near the critical point maximizes dynamic range, information transmission and stimulus discrimination in vivo.

Emergent, collective information processing: Peak information transmission and dynamic range are reduced in cortical networks when synaptic transmission is disrupted. These functions are therefore a consequence of how neurons interact with each other. Signatures pointing toward criticality are coincident with peaks in information processing. Optimal information processing for these functions is thus a collective phenomenon that emerges near the critical point.

Universality: We know that many different species follow the same exponent relation with $\gamma \approx 1.3$, even though they may differ in the number of cortical layers they have or in the cell types and microcircuitry of their brains.

Health: Multiple studies indicate that operating away from the critical point is associated with health problems like epilepsy, sleep deprivation, and Alzheimer's disease.

Homeostasis: After perturbations push the cortex away from the critical point, there is evidence that homeostatic processes bring it back. This has been demonstrated at multiple time scales and in multiple settings, using in vitro preparations, in vivo preparations, and humans.

Not all brain regions near critical: Multiple studies using LFP and spike data demonstrate that layer 2/3 cortical neurons have signatures of operating near the critical point, while layer 5 neurons show little evidence of this.

What We Don't Know

Here, I list questions that have been uncovered by criticality research but are presently unanswered. I expect these areas will experience the most active research in the coming years.

If not critical, then what? The branching ratio is not exactly at the critical value of 1, the susceptibility curves do not diverge to infinity, and the effective exponents drift under varying conditions. All this makes it very unlikely that the cortex is operating exactly *at* the critical point. At the same time, there is overwhelming evidence that it is often operating *near* the critical point. This then raises the question of how to properly characterize what is going on. Is the cortex quasicritical, slightly subcritical or in a Griffiths phase? Or are we merely unable to tell because we are subsampling it too much? Perhaps there is some yet undiscovered way to characterize this that will be more accurate.

What type of phase transition? The branching model we discussed has an inactive to active phase transition. While this was good for didactic purposes, other types of transitions are also possible in the cortex: non-oscillating to oscillating (Poil et al. 2012), attractive dynamics to chaotic dynamics (Dahmen et al. 2019), disorder to order. There are also proposals that there is not just a critical point, but a critical line or a critical manifold separating phases in higher dimensional space (Kanders et al. 2020; Gross 2021). At present we simply do not know which of these many possibilities best applies to the cortex.

When and why does the cortex depart from being nearly critical? Focused attention in humans causes a shift toward subcritical dynamics (Fagerholm et al. 2015). When a zebrafish larva flicks its tail or responds to a change in sensory input, it briefly deviates from being near the critical point (Ponce-Alvarez et al. 2018). While stimulus discrimination is best in turtles when they operate near the critical point, stimulus detection is not (Clawson et al. 2017). These findings show that being near the critical point is not necessarily best for all tasks. Studies with neuromorphic chips and simulated evolution suggest that complex tasks are best performed near criticality while simple tasks are not. But this is poorly understood at present, and much work could be done here. When first learned, many tasks appear to be complex. After much practice, they subjectively appear simple. Related to this, we do not know if there is a transfer of representations from nearly critical regions to less critical regions as learning progresses. Is the critical point useful for the complex and unexpected, but a hinderance for routine situations?

How is being near the critical point related to human cognition? There are studies claiming that proximity to the critical point correlates positively with fluid intelligence (Ezaki et al. 2020; Xu, Yu, and Feng 2021). This fascinating work is still very new, and it will be interesting to see if future studies continue to corroborate it.

Is there a universal set of critical exponents for cortex in the limit of no external drive? If so, then the concept of universality that has been so well established in physics would be applicable to neuroscience. If not, then it could reveal a need for new theories or a broader understanding of the critical point and phase transitions. Perhaps biology could drive physics here.

What are the mechanisms of homeostasis toward the critical point? It is highly likely that synaptic scaling and firing rate homeostasis are involved at the cortical circuit level, but there are details that we do not yet understand. Why does critical homeostasis occur before firing rate homeostasis? What is the role of inhibitory neurons? What types of connectivity patterns promote critical homeostasis? Moving to larger levels, we do not know how lack of sleep causes a departure from the critical point, or how restorative sleep brings it back. The homeostatic mechanisms operating during sensory adaptation must be quick, while those operating over development have longer time scales. Is there overlap between these mechanisms, or are there many distinct processes with minimal overlap? Computational models will likely provide many hypotheses, but only experiments can show which are correct.

Do failures to regulate criticality lead to neurological disorders or is it the other way around? Our lack of knowledge about the mechanisms of critical homeostasis prevents us from answering this chicken-and-egg question. It is also plausible that there are intertwined feedback loops so that no single causative factor can be isolated.

Why are some brain regions close to the critical point while others are not? I gave a hypothesis about why this might be in chapter 8, but this is untested and other hypotheses are possible. We have very little solid knowledge in this area now, and it is open for investigation.

What is the relationship between behavior and criticality? Various reports have noted that human and animal behavior is bursty and may follow scale-free distributions (Barabasi 2005; Kello et al. 2010; Proekt et al. 2012). Are these behaviors driven by nearly critical activity in the brain? And do such behaviors drive the brain closer to the critical point?

Is criticality related to consciousness? We know that criticality is related to, but not identical with, a measure called neural complexity (Tononi, Sporns, and Edelman 1994; Timme et al. 2016). Neural complexity has been proposed to indicate the level of conscious awareness (Tononi and Edelman 1998) and declines with anesthesia or sleep. A growing literature is probing these possible connections (Alonso et al. 2014; Solovey et al. 2015; Tagliazucchi et al. 2016; Varley et al. 2020). While there are tantalizing suggestions here, we do not know enough yet and this area is near the frontier of the field.

Frontier Issues

These topics are primarily outside of neuroscience but are very likely related to the concept of criticality in the brain. As such, progress in these areas may inform neuroscience research into the critical point.

Synthetic systems operating near criticality: There is evidence from multiple labs that networks of nanowires or atomic switches can optimally process information and learn when they are organized to operate near the critical point (Stieg et al. 2012; Pike et al. 2020; Shirai et al. 2020; Hochstetter et al. 2021). There, they produce avalanches that follow the exponent relation and show avalanche shape collapse. Will artificial neural networks in deep learning also benefit from being explicitly constructed to operate near the critical point?

Diverse biological systems are nearly critical: Why is there evidence pointing toward criticality in flocks of birds (Cavagna et al. 2010), the cochlea (Magnasco 2003), the immune system (Mora et al. 2010), networks of genes (Kauffman 1969; Shmulevich, Kauffman, and Aldana 2005), cellular membranes (Veatch et al. 2008) and even single protein molecules (Vattay et al. 2015; Zhang et al. 2017)? Are these examples of convergent evolution?

The role of the environment in criticality: We know that natural images (Ruderman and Bialek 1994; Saremi and Sejnowski 2013) and sounds (Attias 1997) exhibit scale-free statistics. What role does such an environment play in bringing brains and other biological systems toward criticality? Are we living in an environment where being scale free is most adaptive (Stringer et al. 2019)? Are the laws of physics themselves in some sense critical, creating a complex world that is poised between boring order and random chaos?

What I Did Not Cover

The arc of this book has come to an end. In reflecting back on it, I see more clearly that I have taken a narrow path in what is quickly becoming a broad field. I chose to focus on the branching model, when there is a host of other models that probably could have served just as well. I mostly drew from spiking data at a time when LFP, MEG, and fMRI data are increasingly used. I spent nearly all the discussion on what was happening in mammalian cortex, though there is growing evidence that the concept of criticality applies beyond this. While I have tried to be as objective as possible, my presentation inevitably reflects my biases about what is "important" in the field and my recollections about how it developed. As I said in the Introduction, my hope was that by sticking to one model to explain one data type in one brain area, I could keep the story simple and understandable. This approach also kept me close to what I know best, hopefully reducing mistakes.

To ensure that readers are somewhat aware of what else can be found out there, I now want to take a few moments to mention some of the excellent work that I did not have space to explain.

Dante Chialvo has been a pioneer in this field and has been an inspiration to many, including myself. He and Per Bak developed models predicting that the brain would operate near the critical point, years before experiments suggested this to be true (Chialvo and Bak 1999). With colleagues, he has shown from fMRI data that the human brain operates near the critical point (Tagliazucchi et al. 2012). He also showed the most commonly observed patterns of fMRI activity in the human brain can be reproduced by an Ising model tuned to the critical point, when its connections follow those of the human connectome (Haimovici et al. 2013). His overview articles provide excellent perspectives on criticality research in neuroscience (Chialvo 2010, 2004). He is extremely active and a good way to catch up on the latest in this field is just to scan his most recent publications.

Elad Schneideman, Bill Bialek, and colleagues have taken the Ising model and mapped it onto data from spiking neural networks in what they have described as a maximum entropy approach (Schneidman et al. 2006; Tkačik et al. 2009). Like the Ising model, these maximum entropy models have a critical point with functions that show a sharp peak there. This powerful method provides another way to probe the criticality hypothesis (Tkačik et al. 2015). They have recently extended this work to draw connections to the renormalization group, a technique from statistical physics used to determine critical exponents (Meshulam et al. 2019). With Thierry Mora, Bill Bialek has written a highly cited review that surveys criticality research in biology, showing that many systems appear to be poised near the critical point (Mora and Bialek 2011).

Lucilla de Arcangelis has closely analyzed the temporal structure of neuronal avalanches and found that they are significantly correlated in time. Shortly after a given avalanche, a subsequent avalanche is likely to be large, but long after it, a subsequent avalanche is likely to be small (Lombardi et al. 2016). This type of temporal correlation cannot be captured by the bare branching model unless homeostatic mechanisms are added. She has developed models that can reproduce these important temporal correlations (Lombardi et al. 2012). As we saw in the studies by Palva et al. (2013), temporal correlations are intimately tied to the critical point. My presentation spent too little space explaining this essential area, and I hope that readers will look into it as they continue to learn about the critical point.

To these scientists and others I have not mentioned enough, I offer my appreciation for making this a field that has been extremely stimulating and fun to work in.

Appendix

This appendix contains more detailed explanations of material that would have interrupted the flow of the main text. Topics here are organized in the sequence in which they appeared in the text.

Relation between Power-Law Exponent and Slope (Chapters 1 and 6)

The plots below show how a simple function like $y = 1/x$ (or $y = x^{-1}$) appears as a curve in regular coordinates (figure A.1A), but as a straight line when the axes are log-transformed to be in powers of 10 (figure A.1B). These are called log-log coordinates. We typically plot avalanche size and duration distributions in log-log coordinates, where they often approximate straight lines. The slope of the line gives us the exponent.

When the Average Value of a Power Law Diverges and When It Does Not (Chapters 1 and 6)

Often in this book, we would like to know if the average value of a variable (that is distributed as a power law) will be infinite or not. In cases where it is infinite, it is said to

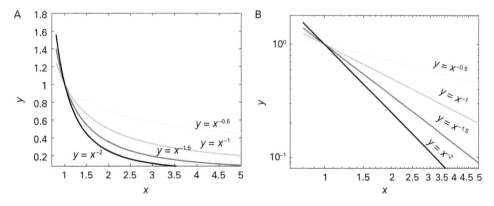

Figure A.1
Power laws in regular coordinates and in log-log coordinates. *A*, Curves for $y = x^{-k}$, where $k = 0.5, 1, 1.5, 2$, plotted for values of x from 0.7 to 5. All curves fall rapidly and asymptotically approach the x-axis as y approaches infinity. *B*, When these equations are plotted in coordinates where the x- and y-axes are in powers of 10, they appear as straight lines. The slope of each line is given by the exponents: $-0.5, -1, -1.5, -2$.

diverge. When it diverges, we can have communication across all scales of the system. It turns out that the exponent of the power law can tell us if it will diverge or not. Here we explain why the exponent can give us this information.

Suppose an avalanche size distribution has the form of:

$$P(S) = kS^{-\tau},$$

where $P(S)$ is the probability of observing an avalanche of size S, k is a proportionality constant, and τ is the exponent of the power law. What would be the average avalanche size in this situation? To calculate the average avalanche size $\langle S \rangle$, also known as the expected value, we will have to multiply the size of the avalanche, S, by its probability of occurrence, $P(S)$ over all sizes. This is done in the following integral:

$$\langle S \rangle = \int_{S=1}^{S=\infty} S \cdot P(S) dS.$$

We will now replace $P(S)$ by the power law:

$$\langle S \rangle = \int_{S=1}^{S=\infty} S \cdot kS^{-\tau} \, dS,$$

which gives

$$\langle S \rangle = k \int_{S=1}^{S=\infty} S^{(1-\tau)} \, dS = k \frac{S^{(2-\tau)}}{(2-\tau)} \Big|_{S=1}^{S=\infty} = k \frac{\infty^{(2-\tau)}}{(2-\tau)} - k \frac{1^{(2-\tau)}}{(2-\tau)} = \frac{k}{(2-\tau)} \left(\infty^{(2-\tau)} - 1^{(2-\tau)} \right).$$

When $\tau = 2$, this will become infinite because the term $\dfrac{k}{(2-\tau)}$ will have zero in the denominator; when $\tau < 2$, the term $(\infty^{(2-\tau)} - 1^{(2-\tau)})$ will become infinite because it will contain infinity raised to some power greater than zero. This means that an exponent of 2 is the dividing line between expected values that are infinite and finite. Note that for a fully connected, infinite critical branching model, the exponent for avalanche durations will be $\alpha = 2$, and for avalanche sizes will be $\tau = 1.5$; these both have expected values that diverge.

Because the mean values of power-law distributions (with exponents less than or equal to 2) are infinite, we should not use the mean to characterize the distribution. Rather, deviations from the mean over a shorter interval, known as fluctuations, become more important to describe it. This leads to analysis measures of long-range temporal correlations (LRTCs), which we describe next.

Long-Range Temporal Correlations (Chapters 1, 6, and 8)

Here we will draw on the excellent descriptions given by Hardstone et al. (2012) and Ihlen (2012) in their review papers.

We are accustomed to characterizing distributions by their means. Teachers do this when they tell us an average score on a test, for example. As we saw above, though, power-law distributions often have a mean that is undefined. In these cases, it is more helpful to characterize the fluctuations of a distribution. For example, if we had data for a stock price over time, we could look at its standard deviation each week and make a distribution from that. It turns out that the standard deviation of such data often grows with the length of the data sampled, allowing us to find a scaling relationship. This can be useful in describing how events are correlated over time. When a system is operating near the critical point, we expect it to have long-range temporal correlations (LRTCs), so measuring these can be informative about the status of the system.

To get an idea of what temporal scaling looks like, let us examine a random walk. We will start our walker at position $x_1 = 0$, at the origin. Each step will be given by a random number drawn from a Gaussian distribution centered on 0 with a standard deviation of 1. Figure A.2 shows the data from one run, displayed for different lengths of time.

From these three simple plots, we can already suspect that there is some scaling involved because they look roughly the same, making it hard to tell if we are zoomed in or zoomed out. This is somewhat surprising, as the last plot is zoomed out by a factor of 10,000 compared to the first plot. In other words, the random walk has the appearance of being scale-free. This is not typical of most objects we see, like a piece of bread, which change dramatically in appearance when we zoom in on them.

To quantify this scaling, let us say we have data, x, sampled at different times, so we have $(x_1, x_2, x_3, \ldots x_t)$. To measure the spread of the data over a given interval, t, we can take its standard deviation (STD). We can compare the standard deviation from an interval t to the standard deviation from an interval Lt, where L is some window length, say 100, by which we multiply the interval t. If there is a scaling relationship in the STD values, we would expect this:

$$STD(Lt) = L^H STD(t).$$

In words, the STD of interval Lt should be equal to the STD of interval t, when it is multiplied by L raised to some exponent H. This exponent H is called the Hurst exponent and can be very informative about the type of temporal scaling we have.

To make this more concrete, let us plug in some example numbers for a random walk that has a mean of zero. Say we run a simulation of a walk for three intervals, each increased by a factor of 100: 1,000 time steps, 100,000 time steps, and 10,000,000 time steps. Since

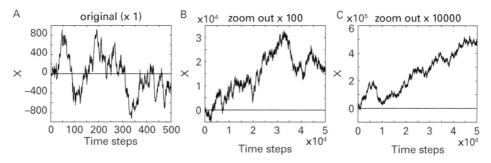

Figure A.2
A random walker plotted at three different time scales. The position of the walker is given by X, on the y-axis, and time is along the x-axis. *A*, 500 time steps. *B*, 50,000 time steps. *C*, 5,000,000 time steps. For each successive plot, the time scale is increased by a factor of 100.

we increase by a factor of 100 each time, this will be our window length L ($L = 100$). Let us say for the sake of example that we measured the standard deviation when the random walk that was 1,000 time steps long and found it to be 32. Also suppose we measured the standard deviation for the random walk after 100,000 time steps long and found it to be 325. For a walk of 10,000,000 we get 3,162. Now let us try to estimate what H would be. We would have:

$$STD(Lt) = L^H STD(t)$$
$$STD(100 \times 1,000) = 100^H STD(1,000)$$
$$325 = 100^H 32$$
$$\frac{325}{32} = 100^H$$
$$10.16 = 100^H.$$

Taking the log (base 10) of both sides, we see that $H \approx 0.5$. This would also be the case if we compared the middle duration with the longest duration samples:

$$STD(Lt) = L^H STD(t)$$
$$STD(100 \times 100,000) = 100^H STD(100,000)$$
$$3,162 = 100^H 325$$
$$\frac{3,162}{325} = 100^H$$
$$9.93 = 100^H.$$

Again, we have $H \approx 0.5$. Thus, a Hurst exponent of 0.5 can tell us how the STD scales with the length of data sampled from a random walk. Note that with a random walk, each step is chosen independently of past steps. Because of this, we say a random walk has no memory.

Now that we have illustrated with a simple example, we can speak in more detail about how to assess scaling in actual data. Very often data will have a strong underlying trend that could be caused by drift. It is helpful to remove such trends by transforming the data into what is called the signal profile, Z. We do this by subtracting the mean from the data and then taking its cumulative sum:

$$Z(t) = \sum_{i=1}^{t} x_i - mean(x).$$

We can then examine the standard deviation of Z to see if this has a scaling relationship. We call the standard deviation of the signal profile the fluctuation function, F:

$$F(t) = STD(Z(t)).$$

As before, we are looking for this relationship:

$$F(Lt) = L^H F(t),$$

where H will be an estimate of the Hurst exponent.

While the random walk has no memory, there are processes that do have memory. For example, consider a random walk where each new flip of the coin would be slightly biased to be like the last flip (i.e., if you got heads last time, your odds of getting heads again was now increased). A plot of this shows how it tends to get increasingly distant from the time axis (figure A.3A). This would be a *correlated* random walk and it would have a Hurst exponent greater than 0.5, say 0.75 (figure A.3B). In a similar manner, we could have an *anticorrelated* random walk, where there would be a slight tendency to reverse direction more often than chance. If you flipped a head, the odds of getting tails next time would be slightly greater than 0.5. An anticorrelated process would stay very close to the time axis (figure A.3A). Here, the Hurst exponent would be less than 0.5 (figure A.3B).

To review:

- A random process with no memory will have a Hurst exponent of 0.5 ($H \approx 0.5$).

- An anticorrelated process will have a Hurst exponent less than 0.5 ($0 \leq H < 0.5$).

- A correlated process will have a Hurst exponent greater than 0.5 ($H > 0.5$).

Recall that when a neural network model is near the critical point, the information from a stimulus will die out more slowly than when the network is subcritical or supercritical (figures 1.4, 1.6). Right at the critical point, the decay of information over time should follow a power law, and this was shown in the seminal paper by Linkenkaer-Hansen et al. (2001) (figure 1.10). This means that long-range temporal correlations can be useful for assessing the state of a network. We would therefore expect the Hurst exponent to be correlated with other measures indicating proximity to the critical point. In the text we review one such study (Palva et al. 2013) that relates exponents from long-range temporal correlations in brain signals to exponents from long-range temporal correlations in behavior (figure 6.8). These exponents are in turn related to exponents from neuronal avalanches.

Although we do not have space to explain here, a closely related form of analysis for long-range temporal correlations is detrended fluctuation analysis, or DFA. In this analysis, trends caused by drift are removed at many scales, leading to a powerful tool for assessing

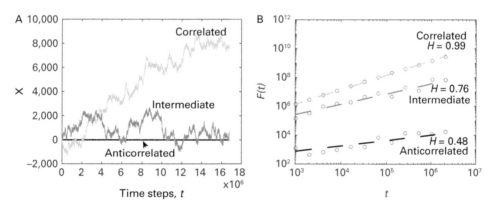

Figure A.3
Temporal scaling of three different types of processes. The processes are correlated, anticorrelated and intermediate. *A*, Each process plotted for 10,000 time steps. Note how the anticorrelated process stays very close to the axis and at this scale is like a line; the correlated process has the largest excursions; the intermediate process has large excursions but comes back to the axis. *B*, The fluctuations $F(t)$ for each process plotted against time, *t*. Axes are in powers of 10, so the straight lines are evidence of scaling. The estimated Hurst exponents are just the slopes of the lines.

temporal scaling. An excellent review of this process is given in Hardstone et al. (2012), and Matlab software for conducting these analyses can be found in Ihlen (2012).

Informal Derivation of the Exponent Relation (Chapters 3, 5, 6, 7, and 8)

Here we describe the reasoning behind the exponent relation. This is an intuitive and informal description, based on the work presented in Scarpetta et al. (2018). What we call the exponent relation has also been called a "scaling relation" or the "crackling noise relation" as this was described in the classic review paper on crackling noise phenomena by Sethna, Dahmen, and Myers (2001). This is the exponent relation:

$$\frac{\alpha - 1}{\tau - 1} = \gamma,$$

where α is the exponent from the avalanche duration distribution (figure A.4A), τ is the exponent from the avalanche size distribution (figure A.4B), and γ is the exponent in the plot of average avalanche size for a given duration, $(\langle S \rangle \,|\, T)$ plotted against duration, T (figure A.4C). We are assuming we have a system that produces power laws for these three relationships. If that is not the case, then the exponent relation will certainly not hold. Additionally, if these distributions are taken from actual data, they will have cutoffs, represented by T_{cutoff} and S_{cutoff}, as shown in the figures.

From figure A.4A, we can say that the probability of observing an avalanche of duration T, which is $P(T)$, scales as the duration T raised to the power $-\alpha$:

$$P(T) \sim T^{-\alpha}.$$

From this, we can say the probability of finding an avalanche of duration $T > T'$ is given by the area under the curve shaded in gray. This area is found by taking the integral of the distribution:

$$P(T > T') = \int_{T'}^{T_{\text{cutoff}}} P(T)\,dt,$$

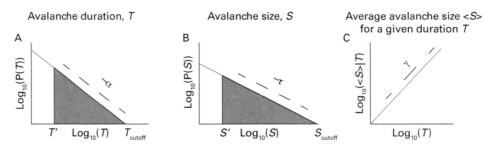

Figure A.4
Understanding the exponent relation. *A*, Schematic of an avalanche duration distribution in log-log coordinates. We assume an exponent of α and that the data has a cutoff duration, T_{cutoff}. The shaded triangle is the area under the curve from some duration T' to the cutoff duration. *B*, Schematic of an avalanche size distribution with exponent τ and a cutoff size S_{cutoff}. *C*, Schematic of average avalanche size for a given duration, $\langle S \rangle \,|\, T$, plotted against duration T.

which will be:

$$P(T > T') = \int_{T'}^{T_{cutoff}} T^{-\alpha} dt = \frac{T^{(1-\alpha)}}{(1-\alpha)} + constant.$$

Where we get a constant of integration, which we can ignore if we are only concerned with how things scale with each other. This means the probability of finding an avalanche of duration $T > T'$ scales as $T^{(1-\alpha)}$, and is given by:

$$P(T > T') \sim T^{(1-\alpha)}.$$

The same reasoning, using an integral, will apply to the avalanche size distribution, and gives us this relationship:

$$P(S > S') \sim T^{(1-\tau)}.$$

If S' is the average avalanche size of an avalanche of duration T', then the probability of finding an avalanche with size larger than S' scales with the probability of finding an avalanche with duration larger than T'. This means just the size of the gray triangles in figure A.4A, B have a scaling relationship. In equation form, this is:

$$P(S > S') \sim P(T > T') \ (*).$$

Now if we substitute the earlier scaling equations for size and duration in for these probabilities, we have:

$$S'^{(1-\tau)} \sim T'^{(1-\alpha)}.$$

The next step is to pull in the other power law relationship for average size and duration, involving the exponent γ. We know empirically that:

$$S' \sim T'^{\gamma}.$$

And we can substitute this in for S' in the equation above it so we have:

$$T'^{(\gamma(1-\tau))} \sim T'^{(1-\alpha)}.$$

Equating the exponents on both sides, we have:

$$\gamma(1-\tau) = (1-\alpha)$$

and with a little algebra we get:

$$\gamma = \frac{(\alpha - 1)}{(\tau - 1)},$$

which is the exponent relation. To arrive at this relationship, we needed the three power laws shown in figure A.4. We also had to assume that equation (*) was true. This would only be the case if the avalanches were fractal copies of each other, and this would only occur if the system was near the critical point. Thus, this exponent relationship can serve as an indicator of proximity to the critical point. It will not be true if the system is far away from the critical point.

Avalanche Shape Collapse (Chapters 3, 5, 6, and 8)

Near the critical point, everything becomes scale-free—this should include average avalanche shapes. To check if avalanche shapes are just rescaled versions of each other, we can try to collapse them. For this process, we need several average avalanche shapes of different sizes (figure A.5A).

The first step is to rescale their durations. We do this by dividing the time in each avalanche by the duration of the avalanche. For example, if an avalanche had events at times [1, 2, 3, 4, 5, 6, 7, 8], we would divide each time by the maximum time (the duration, T) of 8. We would then have these avalanche times [1/8, 2/8, 3/8, 4/8, 5/8, 6/8, 7/8, 8/8], which would be [0.125, 0.250, 0.375, 0.500, 0.625, 0.750, 0.875, 1.00]. Once time is rescaled, we should have something that looks like figure A.3B.

The next step is to rescale their heights. If the total size of the avalanche scales with its height times its duration, then we have:

$$S \sim height \times duration.$$

But we know that the size of the avalanche, S, also scales with the duration raised to the γ power:

$$S \sim T^\gamma.$$

We can use this to make a substitution in the previous equation to get:

$$T^\gamma \sim height \times T.$$

Note that we also substituted in T for duration. If we divide both sides by T, we get:

$$\frac{T^\gamma}{T} \sim height.$$

And this is just:

$$T^{(\gamma-1)} \sim height.$$

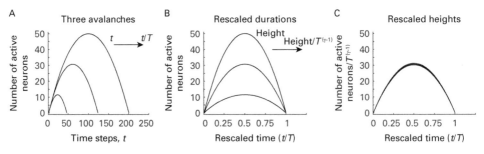

Figure A.5
How to collapse avalanche shapes. *A*, Consider three avalanches of different durations, *T*, and different heights (maximum number of active neurons). We show inverted parabolas, as this shape is often found in neuronal avalanche data. *B*, Rescaling of width occurs when the times, *t*, of activity in each avalanche are divided by the duration, *T*, of each avalanche. This makes them all have the same width. *C*, To rescale their heights, each height is divided by the avalanche duration raised to the ($\gamma - 1$) power. If the avalanches are just fractal copies of each other, as expected near the critical point, there will be a good collapse, as shown here, where the shapes lie nearly on top of each other.

This means that if we divide the height by $T^{(\gamma-1)}$ (or multiply it by its inverse $T^{(1-\gamma)}$) we will be able to rescale the heights so that they are all the same, if we have fractal copies of avalanches. This is shown in figure A.5C.

How to Quantify Network Dynamics (Chapters 4 and 8)

How can we measure dynamical properties in the branching model? In overview, the activity from multiple avalanches can be compared for similarity across time, and as this similarity grows or declines, we can infer trajectories that flow together or apart, respectively. Recall that the activity of a network can be plotted in raster form; avalanches are sequences of consecutively active frames of activity, bracketed by no activity at the beginning and end (figure A.6). Example avalanches of the same lengths from this raster occasionally share active neurons at the same time step. This overlap will form the basis of a similarity measure. This procedure can be conducted for groups of avalanches that have already been found to be significantly similar to each other. See Beggs and Plenz (2004) for a description of how to extract families of similar avalanches from data; software for this is also provided in links later in this appendix.

An easy way to quantify similarity is through the Jaccard coefficient, which is just the number of neurons in the intersection of two patterns, divided by the number of neurons in the union of two patterns. Some example calculations are shown in figure A.7. The Jaccard coefficient can range from 1 (maximal similarity) to 0 (minimal similarity). This allows us to define distance as just 1 minus the similarity. With this, we can measure the distance between two avalanches at every time step, and see if the distance shrinks, grows or stays constant as a function of time.

The three types of dynamics look very different in terms of groups of avalanches (figure A.8). Attractive dynamics are characterized by avalanches that start from widely differing patterns that become increasingly similar or even identical. In contrast, chaotic

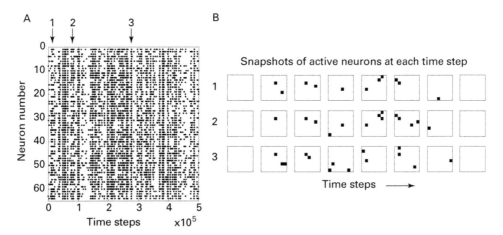

Figure A.6
Avalanches of the same length. *A*, Raster plot of activity from a recurrent branching model with 64 neurons arranged on an 8×8 sheet. Neuron number is along y-axis; time steps along x-axis. *B*, Three example avalanches taken from times indicated by the arrows in panel A. Active neurons indicated by small black squares. (Recall figure 1.13C where avalanche sequences were extracted from the raster plot of LFP data.)

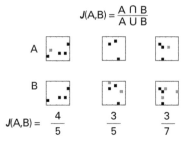

Figure A.7
Measuring similarity. The Jaccard coefficient can quantify the similarity of two patterns of activity, A and B. It is calculated as the number of neurons in the intersection of the two patterns, divided by the number of neurons in the union of the two patterns. In each pair, neurons shared are in black, while those not shared are in gray. Distance between patterns can be defined as 1 minus this similarity.

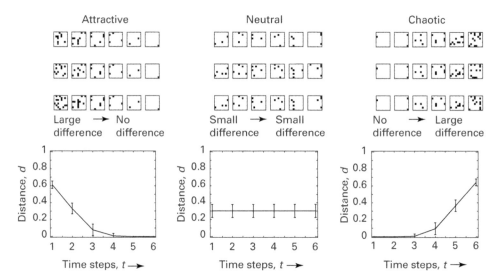

Figure A.8
Identifying dynamics from avalanches. *Left column*, upper panel shows attractive dynamics, where three avalanches start out with dissimilar initial frames, but end with identical frames. Lower panel shows average distance between patterns for each time step for all avalanches from this model. Error bars are standard deviations. *Middle column*, upper panel shows neutral dynamics, where three avalanches have small differences in frames preserved over time. Lower panel shows a constant average distance over time. *Right column*, upper panel shows chaotic dynamics where avalanches start out with identical frames but end up with very large differences. Lower panel shows distance grows over time.

dynamics start with similar patterns but progress toward widely differing patterns. Neutral dynamics have patterns that are roughly equidistant over time.

There is one more step to our characterization of dynamics in these avalanches. In addition to knowing whether avalanches are flowing together or apart, it is useful to measure the rate at which they do so. To capture this rate, we can fit an exponential curve, either declining, growing, or neither, to the distance measurements over time. We can do this repeatedly, for all avalanches produced by the network, to get an average exponential curve. This exponent is called the Lyapunov exponent, Λ, and quantifies the rate of divergence of trajectories. This process of repeated measurement for the three different types of dynamics is schematically drawn in figure A.9.

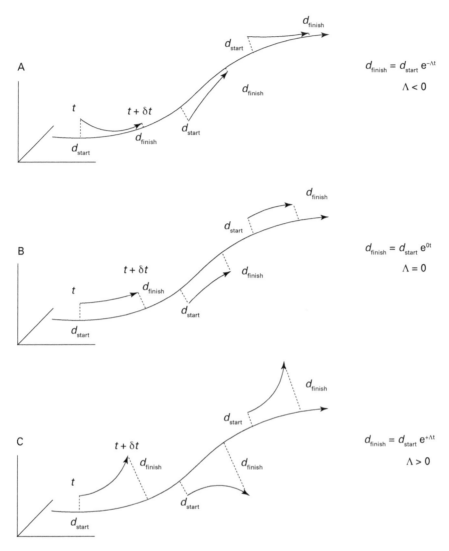

Figure A.9
The Lyapunov exponent gives the rate of attraction or divergence. A, Converging trajectories. An average trajectory in similarity space is shown schematically as a curved line with an arrowhead. When other trajectories are started nearby, at distance d_{start} from the original trajectory, we can measure another distance, d_{finish}, after a short interval δt. The exponent Λ of the best-fitting average curve is called the Lyapunov exponent. Attractive dynamics have $\Lambda < 0$, as distances shrink over time. B, Neutral trajectories preserve distances and have $\Lambda \approx 0$. C, In chaotic dynamics, trajectories grow and $\Lambda > 0$.

As you might expect, the constructed branching ratio of a model network can determine the type of dynamics it will have. When the branching ratio is subcritical, dynamics are attractive; when it is nearly critical, dynamics are approximately neutral; when it is super-critical, dynamics are chaotic. These results are shown for a branching model in figure A.10, in a plot that is very similar to what Haldeman and Beggs (2005) first reported. These models predict that actual cortical networks should be most controllable when they operate near the critical point, a prediction that can be tested experimentally. The methods described here are based on previous work by Derrida and Pomeau (1986), who developed a way to estimate the Lyapunov exponent in random Boolean networks.

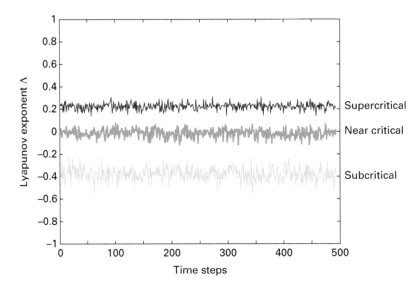

Figure A.10
Near-critical networks have neutral dynamics. Lyapunov exponents, Λ, measured over many time steps for three models with branching ratios of 0.6, 1.0, and 1.4. These models produce negative, near 0, and positive Λ values, respectively. Networks near the critical point are neither attractive nor chaotic and are therefore easier to control.

Software and Data for Exercises and Analyses

Link to Software for Performing Exercises in This Book
These Matlab codes will be needed for the exercises given at the end of some of the chapters. A description of what the codes do, how to use them, and the exercises for each chapter is given here: https://mitpress.mit.edu/cortex-critical-point.

Links to Analysis Toolboxes
The Neural Complexity and Criticality (NCC) toolbox is written in Matlab and described in Marshall et al. (2016), which gives directions for how to use the toolbox. This toolbox contains code for assessing whether a distribution is best fit by a power law or not, and for performing avalanche shape collapse. It also has code for assessing a quantity called neural complexity, which peaks near the critical point. Here is a link to a zipped folder that contains the toolbox and a Readme file about the code: https://nebula.wsimg.com/98697c7 91c08ce110c0ebe10a6142deb?AccessKeyId=AE70CEF69B3CE16C32E9&disposition =0&alloworigin=1.

The Powerlaw toolbox, described in Alstott, Bullmore, and Plenz (2014), is written in Python and can be used to assess whether a distribution is best fit by a power law or not. Source code and Windows installers of Powerlaw are available from the Python Package Index, PyPI, at https://pypi.python.org/pypi/powerlaw. The toolbox can be readily installed with pip using this command: pip install powerlaw. Source code is also available on GitHub at https://github.com/jeffalstott/powerlaw and Google Code at https://code.google.com/p /powerlaw/.

A Matlab toolbox for assessing long-range temporal correlations is described in Ihlen (2012), and can be found here: https://www.ntnu.edu/inb/geri/software.

Another Matlab toolbox for long-range temporal correlations is described in Hardstone et al. (2012). This Neurophysiological Biomarker Toolbox (NBT) gives practical advice, a tutorial and can be found here: https://www.frontiersin.org/articles/10.3389/fphys.2012 .00450/full#h6.

Publicly Available Data Sets
Spike data from cortical slice cultures, recorded in the author's lab on a 512-electrode array: Shinya Ito, Fang-Chin Yeh, Nicholas M. Timme, Pawel Hottowy, Alan M. Litke, and John M. Beggs (2016); Spontaneous spiking activity of hundreds of neurons in mouse somatosensory cortex slice cultures recorded using a dense 512-electrode array. http://dx .doi.org/10.6080/K07D2S2F.

Spike data from dissociated cultures, recorded in the author's lab on a 60-electrode array: Nicholas M. Timme, Najja Marshall, Nicholas Bennett, Monica Ripp, Edward Lautzenhiser, and John M. Beggs (2016); Spontaneous spiking activity of thousands of neurons in rat hippocampal dissociated cultures. http://dx.doi.org/10.6080/K0PC308P.

Freely available data, including spikes and LFPs from many brain regions from many labs: These are posted at the Collaborative Research in Computational Neuroscience (CRCNS) site. https://crcns.org/.

Notes

Chapter 1

1. This same basic idea holds for other types of information transmission such as which neurons had been previously active. A more detailed description of information will be given in chapter 4.

2. A reflected random walk can produce power laws. Reflection occurs whenever the random walk hits the time axis from above, so that the resulting plot never has negative values. The sizes and durations of the excursions from the axis are power-law distributed (Kostinski and Amir, 2016).

Chapter 2

1. Indeed, more quantitative studies have shown that in some cases, it is impossible to deduce the macro states from even complete knowledge of the micro rules: Gu et al. (2009) and Cubitt, Perez-Garcia, and Wolf (2015).

Chapter 3

1. For those who like a more physics-based explanation: In thermodynamics, first-order phase transitions show a discontinuity in the first derivative of the free energy, while second-order phase transitions show a discontinuity in the second derivative of the free energy.

Chapter 4

1. The mutual information is actually bidirectional, and the output contains just as much information about the input as the input does about the output.

2. To be more precise, they defined it as $\Delta = log_{10} \dfrac{Response_{90}}{Response_{10}}$, which is just the logarithm of the difference between the response strength at the 90th percentile and the response strength at the 10th percentile.

3. Briefly, when very few samples are taken, the entropy of distributions is underestimated and this biases the mutual information measures.

Chapter 5

1. The specific heat is related to the amount by which the energy changes if we were to flip one spin.

Chapter 6

1. STDP is an acronym for spike timing-dependent plasticity. In its most common form, when neuron A consistently sends a synaptic impulse (20 milliseconds or less) before neuron B fires a spike, the connection between neuron A and B will be strengthened. When neuron A's impulse consistently arrives (1 to 20 milliseconds) after neuron B spikes, the connection will be weakened. This is a timing-dependent version of Hebb's famous synaptic plasticity rule (Hebb 2005).

2. To specify which of the ~86 billion neurons to connect to, you would need $log_2(86 \times 10^9) = 36$ bits for each synapse. There are about 86 billion $\times$ 1,000 synapses. The total information needed then would be $36 \times 86 \times 10^9 \times 1,000 = 3.1 \times 10^{15}$ bits. And 3.1×10^{15} bits divided by 700 MB gives about 4.4 billion CDs.

Chapter 7

1. This fact suggests that nonequilibrium models are the most appropriate for cortical networks. While equilibrium models can serve as approximations, they cannot capture the details of changing external drive and avalanche dynamics unfolding over time.

2. As we will see, critical exponents only exist in the limit of zero external drive. To recognize this fact, the exponents that occur for nonzero external drive are called "effective exponents."

3. The word *dynamical* is meant to signify that this is a nonequilibrium model, constantly receiving external inputs. This contrasts with an equilibrium model which would be a closed system with no external inputs. There, we would have static susceptibility.

4. Ben Widom was my freshman-year chemistry professor at Cornell. About 33 years later, when we published a paper connecting the Widom line to neural network behavior near the critical point (Williams-Garcia et al. 2014), I contacted him. He was amused and stated that the good work was surely a result of my excellent undergraduate education in chemistry.

Chapter 8

1. We do not have the space to discuss interpretations of intelligence or how it should be measured in different species. Here, I only mean it in the most colloquial or generic way—what would we intuitively call smart.

2. Cortex means "bark" in Latin. The cortex is to the brain like the bark is to a tree—its outer covering. Here, we will be primarily talking about neocortex, or "new bark," the most recently evolved regions that have six layers. There is also four- or five-layered paleocortex, "old bark," as well as three-layered archicortex, found in olfactory areas and the hippocampus. Here, when I use the term cortex, I will be referring to neocortex.

3. There are two uses of the term "layer" in the different literatures: in cortical anatomy, a layer refers to a narrow band of cell types of particular connectivity within the cortical thickness; in deep learning, a layer refers to a population of neuron-like processing elements. Deep learning networks became more powerful by adding many more layers (or processing stages), but neocortex in mammals has not deviated much from six basic layers.

4. Even the simplest artificial neural networks have at least three processing stages: an input layer, a hidden layer, and an output layer. If a network is to become deeper, it is only by adding more hidden layers. Here, a single-layered network refers to a network that has only one hidden layer. Deep neural networks may have over a thousand layers (He et al. 2016).

5. Calculating parity merely means to determine whether the sum of all the inputs is even or odd.

References

Abbott, Larry F., and Sacha B. Nelson. 2000. "Synaptic Plasticity: Taming the Beast." *Nature Neuroscience* 3:1178–1183.

Acebrón, Juan A., Luis L. Bonilla, Conrad J. Pérez Vicente, Félix Ritort, and Renato Spigler. 2005. "The Kuramoto Model: A Simple Paradigm for Synchronization Phenomena." *Reviews of Modern Physics* 77:137.

Adelsberger, Helmuth, Olga Garaschuk, and Arthur Konnerth. 2005. "Cortical Calcium Waves in Resting Newborn Mice." *Nature Neuroscience* 8:988–990.

Agrawal, Vidit, Srimoy Chakraborty, Thomas Knöpfel, and Woodrow L. Shew. 2019. "Scale-Change Symmetry in the Rules Governing Neural Systems." *Iscience* 12:121–131.

Aguilera, M., C. Alquézar, and E. J. Izquierdo. 2017. "Signatures of Criticality in a Maximum Entropy Model of the C. Elegans Brain during Free Behaviour." *Artificial Life Conference Proceedings* 14:29–35.

Alonso, Leandro M., Alex Proekt, Theodore H. Schwartz, Kane O. Pryor, Guillermo A. Cecchi, and Marcelo O. Magnasco. 2014. "Dynamical Criticality during Induction of Anesthesia in Human ECoG Recordings." *Frontiers in Neural Circuits* 8:20.

Alstott, Jeff, Ed Bullmore, and Dietmar Plenz. 2014. "Powerlaw: A Python Package for Analysis of Heavy-Tailed Distributions." *PLoS One* 9:e85777.

Andersen, Peter Bøgh, Claus Emmeche, Niels Ole Finnemann, and Peder Voetmann Christiansen. 2000. *Downward Causation*. Aarhus, Denmark: Aarhus University Press.

Anderson, Philip W. 1972. "More Is Different." *Science* 177:393–396.

Arviv, Oshrit, Abraham Goldstein, and Oren Shriki. 2015. "Near-Critical Dynamics in Stimulus-Evoked Activity of the Human Brain and Its Relation to Spontaneous Resting-State Activity." *Journal of Neuroscience* 35:13927–13942.

Arviv, Oshrit, Mordekhay Medvedovsky, Liron Sheintuch, Abraham Goldstein, and Oren Shriki. 2016. "Deviations from Critical Dynamics in Interictal Epileptiform Activity." *Journal of Neuroscience* 36:12276–12292.

Attias, Hagai, and C. E. Schreiner. 1997. "Temporal Low-Order Statistics of Natural Sounds." *Advances in Neural Information Processing Systems* 9:27.

Azevedo, Frederico A. C., Ludmila R. B. Carvalho, Lea T. Grinberg, José Marcelo Farfel, Renata E. L. Ferretti, Renata E. P. Leite, Wilson Jacob Filho, Roberto Lent, and Suzana Herculano-Houzel. 2009. "Equal Numbers of Neuronal and Nonneuronal Cells Make the Human Brain an Isometrically Scaled-Up Primate Brain." *Journal of Comparative Neurology* 513:532–541.

Babloyantz, Agnessa, and Alain Destexhe. 1986. "Low-Dimensional Chaos in an Instance of Epilepsy." *Proceedings of the National Academy of Sciences* 83:3513–3517.

Badre, David, and Mark D'Esposito. 2009. "Is the Rostro-Caudal Axis of the Frontal Lobe Hierarchical?" *Nature Reviews: Neuroscience* 10:659.

Bak, Per. 1996. *How Nature Works: The Science of Self-Organized Criticality*. New York: Copernicus.

Bak, Per, Chao Tang, and Kurt Wiesenfeld. 1987. "Self-Organized Criticality: An Explanation of the 1/f Noise." *Physical Review Letters* 59:381.

Ballerini, Michele, Nicola Cabibbo, Raphael Candelier, Andrea Cavagna, Evaristo Cisbani, Irene Giardina, Vivien Lecomte, Alberto Orlandi, Giorgio Parisi, and Andrea Procaccini. 2008. "Interaction Ruling Animal Collective Behavior Depends on Topological Rather than Metric Distance: Evidence from a Field Study." *Proceedings of the National Academy of Sciences* 105:1232–1237.

Barabasi, Albert-Laszlo. 2005. "The Origin of Bursts and Heavy Tails in Human Dynamics." *Nature* 435:207–211.

Barbas, Helen, and Nancy Rempel-Clower. 1997. "Cortical Structure Predicts the Pattern of Corticocortical Connections." *Cerebral Cortex* 7:635–646.

Barras, Colin. 2013. "Mind Maths: The Equations of Thought." *New Scientist* 35.

Barrett, Lisa Feldman. 2020. *Seven and a Half Lessons about the Brain*. Boston, MA: Houghton Mifflin Harcourt.

Barron, Andrew R. 1994. "Approximation and Estimation Bounds for Artificial Neural Networks." *Machine Learning* 14:115–133.

Bedard, Claude, Helmut Kroeger, and Alain Destexhe. 2006. "Does the 1/f Frequency Scaling of Brain Signals Reflect Self-Organized Critical States?" *Physical Review Letters* 97:118102.

Bedau, Mark. 2002. "Downward Causation and the Autonomy of Weak Emergence." *Principia: An International Journal of Epistemology* 6:5–50.

Beggs, J. M. 2008. "The Criticality Hypothesis: How Local Cortical Networks Might Optimize Information Processing." *Philosophical Transactions. Series A, Mathematical, Physical, and Engineering Sciences* 366:329–343.

Beggs, J. M. 2015. "Can There Be a Physics of the Brain?" *Physical Review Letters* 114 (22): 220001.

Beggs, J. M. 2019. "The Critically Tuned Cortex." *Neuron* 104:623–624.

Beggs, J. M., and D. Plenz. 2003. "Neuronal Avalanches in Neocortical Circuits." *Journal of Neuroscience* 23 (35): 11167–11177.

Beggs, J. M., and D. Plenz. 2004. "Neuronal Avalanches are Diverse and Precise Activity Patterns That Are Stable for Many Hours in Cortical Slice Cultures." *Journal of Neuroscience* 24:5216–5229.

Beggs, J. M., and N. Timme. 2012. "Being Critical of Criticality in the Brain." *Frontiers in Physiology* 3:163.

Begley, Charles E., and Tracy L. Durgin. 2015. "The Direct Cost of Epilepsy in the United States: A Systematic Review of Estimates." *Epilepsia* 56:1376–1387.

Bellay, Timothy, Andreas Klaus, Saurav Seshadri, and Dietmar Plenz. 2015. "Irregular Spiking of Pyramidal Neurons Organizes as Scale-Invariant Neuronal Avalanches in the Awake State." *Elife* 4:e07224.

Benayoun, Marc, Jack D. Cowan, Wim van Drongelen, and Edward Wallace. 2010. "Avalanches in a Stochastic Model of Spiking Neurons." *PLOS Computational Biology* 6:e1000846.

Berche, Bertrand, Malte Henkel, and Ralph Kenna. 2009. "Critical Phenomena: 150 Years Since Cagniard de la Tour." *arXiv Preprint arXiv:0905.1886.*

Berg, Richard E. and David G. Stork, 1995. *The Physics of Sound*, 2nd ed. Englewood Cliffs, NJ: Prentice Hall.

Bertschinger, N., and T. Natschlager. 2004. "Real-Time Computation at the Edge of Chaos in Recurrent Neural Networks." *Neural Computation* 16:1413–1436.

Bi, Guo-qiang, and Mu-ming Poo. 1998. "Synaptic Modifications in Cultured Hippocampal Neurons: Dependence on Spike Timing, Synaptic Strength, and Postsynaptic Cell Type." *Journal of Neuroscience* 18:10464–10472.

Bialek, William, Andrea Cavagna, Irene Giardina, Thierry Mora, Oliver Pohl, Edmondo Silvestri, Massimiliano Viale, and Aleksandra M. Walczak. 2014. "Social Interactions Dominate Speed Control in Poising Natural Flocks Near Criticality." *Proceedings of the National Academy of Sciences* 111:7212–7217.

Bianchini, Monica, and Franco Scarselli. 2014. "On the Complexity of Neural Network Classifiers: A Comparison between Shallow and Deep Architectures." *IEEE Transactions on Neural Networks and Learning Systems* 25:1553–1565.

Bienenstock, Elie. 1995. "A Model of Neocortex." *Network: Computation in Neural Systems* 6:179–224.

Bishop, Robert C. 2008. "Downward Causation in Fluid Convection." *Synthese* 160:229–248.

Blair, Hugh T., Adam C. Welday, and Kechen Zhang. 2007. "Scale-Invariant Memory Representations Emerge from Moire Interference between Grid Fields That Produce Theta Oscillations: A Computational Model." *Journal of Neuroscience* 27:3211–3229.

Bonachela, Juan A., Sebastiano De Franciscis, Joaquín J. Torres, and Miguel A. Muñoz. 2010. "Self-Organization without Conservation: Are Neuronal Avalanches Generically Critical?" *Journal of Statistical Mechanics: Theory and Experiment* 2010:P02015.

Bonachela, Juan A., and Miguel A. Muñoz. 2009. "Self-Organization without Conservation: True or Just Apparent Scale-Invariance?" *Journal of Statistical Mechanics: Theory and Experiment* 2009:P09009.

Bonifazi, Paolo, Miri Goldin, Michel A. Picardo, Isabel Jorquera, A. Cattani, Gregory Bianconi, Alfonso Represa, Yehezkel Ben-Ari, and Rosa Cossart. 2009. "GABAergic Hub Neurons Orchestrate Synchrony in Developing Hippocampal Networks." *Science* 326:1419–1424.

Bowen, Zac, Daniel E. Winkowski, Saurav Seshadri, Dietmar Plenz, and Patrick O. Kanold. 2019. "Neuronal Avalanches in Input and Associative Layers of Auditory Cortex." *Frontiers in Systems Neuroscience* 13:45.

Braitenberg, Valentino, and Almut Schüz. 1998. *Cortex: Statistics and Geometry of Neuronal Connectivity.* Berlin: Springer Verlag.

Breakspear, M., S. Heitmann, and A. Daffertshofer. 2010. "Generative Models of Cortical Oscillations: Neurobiological Implications of the Kuramoto Model." *Frontiers in Human Neuroscience* 4:190.

Breskin, Ilan, Jordi Soriano, Elisha Moses, and Tsvi Tlusty. 2006. "Percolation in Living Neural Networks." *Physical Review Letters* 97:188102.

Bressloff, Paul C. 2014. *Waves in Neural Media.* New York: Springer.

Bretas, Rafael Vieira, Yumiko Yamazaki, and Atsushi Iriki. 2020. "Phase Transitions of Brain Evolution That Produced Human Language and Beyond." *Neuroscience Research* 161:1–7.

Briggman, Kevin L., Henry D. I. Abarbanel, and William B. Kristan. 2005. "Optical Imaging of Neuronal Populations during Decision-Making." *Science* 307:896–901.

Brunel, Nicolas. 2000. "Dynamics of Sparsely Connected Networks of Excitatory and Inhibitory Spiking Neurons." *Journal of Computational Neuroscience* 8:183–208.

Brush, Stephen G. 1967. "History of the Lenz-Ising Model." *Reviews of Modern Physics* 39:883.

Buchman, Timothy G. 2002. "The Community of the Self." *Nature* 420:246–251.

Burgess, Neil, Caswell Barry, and John O'Keefe. 2007. "An Oscillatory Interference Model of Grid Cell Firing." *Hippocampus* 17:801–812.

Buzsáki, György. 2002. "Theta Oscillations in the Hippocampus." *Neuron* 33: 325–340.

Buzsáki, György. 2006. *Rhythms of the Brain.* Oxford: Oxford University Press.

Buzsáki, György, and Andreas Draguhn. 2004. "Neuronal Oscillations in Cortical Networks." *Science* 304: 1926–1929.

Calvin, William H. 1994. "The Emergence of Intelligence." *Scientific American* 271:100–107.

Cappaert, N. L. M., F. H. Lopes da Silva, and W. J. Wadman. 2009. "Spatio-Temporal Dynamics of Theta Oscillations in Hippocampal–Entorhinal Slices." *Hippocampus* 19:1065–1077.

Carandini, Matteo, and David J. Heeger. 2012. "Normalization as a Canonical Neural Computation." *Nature Reviews Neuroscience* 13:51–62.

Carr, C. E., and R. E. Boudreau. 1993. "Organization of the Nucleus Magnocellularis and the Nucleus Laminaris in the Barn Owl: Encoding and Measuring Interaural Time Differences." *Journal of Comparative Neurology* 334:337–355.

Carr, C. E., and M. Konishi. 1990. "A Circuit for Detection of Interaural Time Differences in the Brain Stem of the Barn Owl." *Journal of Neuroscience* 10:3227–3246.

Carrillo-Reid, Luis, Shuting Han, Weijian Yang, Alejandro Akrouh, and Rafael Yuste. 2019. "Controlling Visually Guided Behavior by Holographic Recalling of Cortical Ensembles." *Cell* 178:447–457, e5.

Carvalho, Tawan T. A., Antonio J. Fontenele, Mauricio Girardi-Schappo, Thais Feliciano, Leandro A. A. Aguiar, Thais P. L. Silva, Nivaldo A. P. de Vasconcelos, Pedro V. Carelli, and Mauro Copelli. 2021. "Subsampled Directed-Percolation Models Explain Scaling Relations Experimentally Observed in the Brain." *Frontiers in Neural Circuits* 2021: 83.

Cavagna, Andrea, Alessio Cimarelli, Irene Giardina, Giorgio Parisi, Raffaele Santagati, Fabio Stefanini, and Massimiliano Viale. 2010. "Scale-Free Correlations in Starling Flocks." *Proceedings of the National Academy of Sciences* 107:11865–11870.

Chaudhuri, Rishidev, Kenneth Knoblauch, Marie-Alice Gariel, Henry Kennedy, and Xiao-Jing Wang. 2015. "A Large-Scale Circuit Mechanism for Hierarchical Dynamical Processing in the Primate Cortex." *Neuron* 88:419–431.

Chen, Janice, Uri Hasson, and Christopher J. Honey. 2015. "Processing Timescales as an Organizing Principle for Primate Cortex." *Neuron* 88:244–246.

Chen, W., J. P. Hobbs, A. Tang, and J. M. Beggs. 2010. "A Few Strong Connections: Optimizing Information Retention in Neuronal Avalanches." *BMC Neuroscience* 11:3.

Chialvo, Dante R. 2004. "Critical Brain Networks." *Physica A: Statistical Mechanics and its Applications* 340:756–765.

Chialvo, Dante R. 2010. "Emergent Complex Neural Dynamics." *Nature Physics* 6:744–750.

Chialvo, Dante R., and Per Bak. 1999. "Learning from Mistakes." *Neuroscience* 90:1137–1148.

Chien, Hsiang-Yun Sherry, Javier S. Turek, Nicole Beckage, Vy A. Vo, Christopher J. Honey, and Ted L. Willke. 2021. "Slower is Better: Revisiting the Forgetting Mechanism in LSTM for Slower Information Decay." *arXiv Preprint arXiv:2105.05944.*

Ciresan, Dan, Ueli Meier, Jonathan Masci, and Jurgen Schmidhuber. 2012. "Multi-Column Deep Neural Network for Traffic Sign Classification." *Neural Networks* 32.

Clauset, A., C. R. Shalizi, and M. E. Newman. 2009. "Power-Law Distributions in Empirical Data." *SIAM Review* 51 (4): 661–703.

Clawson, Wesley P., Nathaniel C. Wright, Ralf Wessel, and Woodrow L. Shew. 2017. "Adaptation Towards Scale-Free Dynamics Improves Cortical Stimulus Discrimination at the Cost of Reduced Detection." *PLOS Computational Biology* 13:e1005574.

Cramer, Benjamin, David Stöckel, Markus Kreft, Michael Wibral, Johannes Schemmel, Karlheinz Meier, and Viola Priesemann. 2020. "Control of Criticality and Computation in Spiking Neuromorphic Networks with Plasticity." *Nature Communications* 11:1–11.

Cubitt, Toby S., David Perez-Garcia, and Michael M. Wolf. 2015. "Undecidability of the Spectral Gap." *Nature* 528:207–211.

Curie, Pierre. 1895. *Propriétés magnétiques des corps a diverses températures.* Paris: Gauthier-Villars et fils.

Dahmen, David, Sonja Grün, Markus Diesmann, and Moritz Helias. 2019. "Second Type of Criticality in the Brain Uncovers Rich Multiple-Neuron Dynamics." *Proceedings of the National Academy of Sciences* 116:13051–13060.

Dalla Porta, L., and M. Copelli. 2019. "Modeling Neuronal Avalanches and Long-Range Temporal Correlations at the Emergence of Collective Oscillations: Continuously Varying Exponents Mimic M/EEG Results." *PLoS Computational Biology* 15 (4): e1006924.

Daniels, Bryan C., Christopher J. Ellison, David C. Krakauer, and Jessica C. Flack. 2016. "Quantifying Collectivity." *Current Opinion in Neurobiology* 37:106–113.

Dave, Amish S., and Daniel Margoliash. 2000. "Song Replay during Sleep and Computational Rules for Sensorimotor Vocal Learning." *Science* 290:812–816.

Davis, Zachary W., Lyle Muller, Julio Martinez-Trujillo, Terrence Sejnowski, and John H. Reynolds. 2020. "Spontaneous Travelling Cortical Waves Gate Perception in Behaving Primates." *Nature* 587:432–436.

De Carvalho, Josué X., and Carmen P. C. Prado. 2000. "Self-Organized Criticality in the Olami-Feder-Christensen Model." *Physical Review Letters* 84:4006.Del Papa, B., V. Priesemann, and J. Triesch. 2017. "Criticality Meets Learning: Criticality Signatures in a Self-Organizing Recurrent Neural Network." *PLoS One* 12.

Derrida, Bernard, and Yves Pomeau. 1986. "Random Networks of Automata: A Simple Annealed Approximation." *EPL (Europhysics Letters)* 1:45.

Destexhe, Alain, and Terrence J. Sejnowski. 2009. "The Wilson–Cowan Model, 36 Years Later." *Biological Cybernetics* 101:1–2.

Destexhe, Alain, and Jonathan D. Touboul. 2021. "Is There Sufficient Evidence for Criticality in Cortical Systems?" *eNeuro* 8.

Diamond, Mathew E., Wei Huang, and Ford F. Ebner. 1994. "Laminar Comparison of Somatosensory Cortical Plasticity." *Science* 265:1885–1888.

Dickson, Clayton T., Jacopo Magistretti, Mark Shalinsky, Bassam Hamam, and Angel Alonso. 2000. "Oscillatory Activity in Entorhinal Neurons and Circuits: Mechanisms and Function." *Annals of the New York Academy of Sciences* 911:127–150.

Doi, Atsushi, Masaharu Mizuno, Toshihiko Katafuchi, Hidemasa Furue, Kohei Koga, and Megumu Yoshimura. 2007. "Slow Oscillation of Membrane Currents Mediated by Glutamatergic Inputs of Rat Somatosensory Cortical Neurons: In Vivo Patch-Clamp Analysis." *European Journal of Neuroscience* 26:2565–2575.

Domany, Eytan, and Wolfgang Kinzel. 1984. "Equivalence of Cellular Automata to Ising Models and Directed Percolation." *Physical Review Letters* 53:311.

Douglas, Rodney J., Kevan A. C. Martin, and David Whitteridge. 1989. "A Canonical Microcircuit for Neocortex." *Neural Computation* 1:480–488.

Dunkelmann, S., and G. Radons. 1994. "Neural Networks and Abelian Sandpile Models of Self-Organized Criticality." In *Proceedings of International Conference Artificial Neural Networks* 867–870. Springer-Verlag.

Eckhorn, Reinhard, Roman Bauer, Wolfgang Jordan, Michael Brosch, Wolfgang Kruse, Matthias Munk, and H. J. Reitboeck. 1988. "Coherent Oscillations: A Mechanism of Feature Linking in the Visual Cortex?" *Biological Cybernetics* 60:121–130.

Effenberger, Felix, Jürgen Jost, and Anna Levina. 2015. "Self-Organization in Balanced State Networks by STDP and Homeostatic Plasticity." *PLoS Computational Biology* 11:e1004420.

Eurich, Christian W., J. Michael Herrmann, and Udo A. Ernst. 2002. "Finite-Size Effects of Avalanche Dynamics." *Physical Review E* 66:066137.

Ezaki, Takahiro, Elohim Fonseca Dos Reis, Takamitsu Watanabe, Michiko Sakaki, and Naoki Masuda. 2020. "Closer to Critical Resting-State Neural Dynamics in Individuals with Higher Fluid Intelligence." *Communications Biology* 3:1–9.

Fagerholm, E. D., R. Lorenz, G. Scott, M. Dinov, P. J. Hellyer, N. Mirzaei, C. Leeson, D. W. Carmichael, D. J. Sharp, W. L. Shew, and R. Leech. 2015. "Cascades and Cognitive State: Focused Attention Incurs Subcritical Dynamics." *Journal of Neuroscience* 35 (11): 4626–4634.

Fagerholm, Erik D., Gregory Scott, Woodrow L. Shew, Chenchen Song, Robert Leech, Thomas Knöpfel, and David J. Sharp. 2016. "Cortical Entropy, Mutual Information and Scale-Free Dynamics in Waking Mice." *Cerebral Cortex* 26:3945–3952.

Faisal, A. Aldo, Luc P. J. Selen, and Daniel M. Wolpert. 2008. "Noise in the Nervous System." *Nature Reviews Neuroscience* 9:292–303.

Farkas, Illés, Dirk Helbing, and Tamás Vicsek. 2002. "Mexican Waves in an Excitable Medium." *Nature* 419:131–132.

Favela, Luis H. 2019. "Emergence by Way of Dynamic Interactions." *Southwest Philosophy Review* 35:47–57.

Feldman, Daniel E., and Michael Brecht. 2005. "Map Plasticity in Somatosensory Cortex." *Science* 310:810–815.

Feldman, Jack L., and Christopher A. Del Negro. 2006. "Looking for Inspiration: New Perspectives on Respiratory Rhythm." *Nature Reviews Neuroscience* 7:232–241.

Felleman, Daniel J., and David C. Van Essen. 1991. "Distributed Hierarchical Processing in the Primate Cerebral Cortex." *Cerebral Cortex* 1:1–47.

Firth, Sally I., Chih-Tien Wang, and Marla B. Feller. 2005. "Retinal Waves: Mechanisms and Function in Visual System Development." *Cell Calcium* 37:425–432.

Fisahn, Andre, Fenella G. Pike, Eberhard H. Buhl, and Ole Paulsen. 1998. "Cholinergic Induction of Network Oscillations at 40 Hz in the Hippocampus In Vitro." *Nature* 394:186–189.

Fleming, Donald. 1984. "Walter B. Cannon and Homeostasis." *Social Research* 609–640.

Fontenele, A. J., N. A. P. de Vasconcelos, T. Feliciano, L. A. A. Aguiar, C. Soares-Cunha, B. Coimbra, L. Dalla Porta, S. Ribeiro, A. J. Rodrigues, N. Sousa, P. V. Carelli, and M. Copelli. 2019. "Criticality between Cortical States." *Physical Review Letters* 122:208101.

Fosque, Leandro J., Rashid V. Williams-García, John M. Beggs, and Gerardo Ortiz. 2021. "Evidence for Quasicritical Brain Dynamics." *Physical Review Letters* 126:098101.

Fraiman, Daniel, Pablo Balenzuela, Jennifer Foss, and Dante R. Chialvo. 2009. "Ising-Like Dynamics in Large-Scale Functional Brain Networks." *Physical Review E* 79:061922.

Freeman, Walter J. 1987. "Simulation of Chaotic EEG Patterns with a Dynamic Model of the Olfactory System." *Biological Cybernetics* 56:139–150.

Friedman, Eric J., and Adam S. Landsberg. 2013. "Hierarchical Networks, Power Laws, and Neuronal Avalanches." *Chaos* 23:013135.

Friedman, N., S. Ito, B. A. Brinkman, M. Shimono, R. E. DeVille, K. A. Dahmen, J. M. Beggs, and T. C. Butler. 2012. "Universal Critical Dynamics in High Resolution Neuronal Avalanche Data." *Physical Review Letters* 108 (20): 208102.

Fries, Pascal. 2005. "A Mechanism for Cognitive Dynamics: Neuronal Communication through Neuronal Coherence." *Trends in Cognitive Sciences* 9:474–480.

Fries, Pascal. 2009. "Neuronal Gamma-Band Synchronization as a Fundamental Process in Cortical Computation." *Annual Review of Neuroscience* 32:209–224.

Friston, Karl. 2010. "The Free-Energy Principle: A Unified Brain Theory?" *Nature Reviews Neuroscience* 11:127–138.

Fusi, Stefano, Patrick J. Drew, and Larry F. Abbott. 2005. "Cascade Models of Synaptically Stored Memories." *Neuron* 45:599–611.

Gal, Asaf, Danny Eytan, Avner Wallach, Maya Sandler, Jackie Schiller, and Shimon Marom. 2010. "Dynamics of Excitability over Extended Timescales in Cultured Cortical Neurons." *Journal of Neuroscience* 30:16332–16342.

Galam, Serge. 1997. "Rational Group Decision Making: A Random Field Ising Model at T = 0." *Physica A: Statistical Mechanics and its Applications* 238:66–80.

Garaschuk, Olga, Jennifer Linn, Jens Eilers, and Arthur Konnerth. 2000. "Large-Scale Oscillatory Calcium Waves in the Immature Cortex." *Nature Neuroscience* 3:452–459.

Gardner, Martin. 1970. "Mathematical Games: The Fantastic Combinations of John Conway's New Solitaire Game 'Life.'" *Scientific American* 223:120–123.

Garey, Laurence J. 1999. *Brodmann's "Localisation in the Cerebral Cortex."* World Scientific.

Gautam, S. H., T. T. Hoang, K. McClanahan, S. K. Grady, and W. L. Shew. 2015. "Maximizing Sensory Dynamic Range by Tuning the Cortical State to Criticality." *PLoS Computational Biology* 11:e1004576.

Georgopoulos, Apostolos P., Andrew B. Schwartz, and Ronald E. Kettner. 1986. "Neuronal Population Coding of Movement Direction." *Science* 233:1416–1419.

Girardi-Schappo, Mauricio, Ludmila Brochini, Ariadne A. Costa, Tawan T. A. Carvalho, and Osame Kinouchi. 2020. "Synaptic Balance Due to Homeostatically Self-Organized Quasicritical Dynamics." *Physical Review Research* 2:012042.

Goodhill, Geoffrey J. 2016. "Can Molecular Gradients Wire the Brain?" *Trends in Neurosciences* 39:202–211.

Graner, François, and James A. Glazier. 1992. "Simulation of Biological Cell Sorting Using a Two-Dimensional Extended Potts Model." *Physical Review Letters* 69:2013.

Gray, Charles M., Peter König, Andreas K. Engel, and Wolf Singer. 1989. "Oscillatory Responses in Cat Visual Cortex Exhibit Inter-Columnar Synchronization Which Reflects Global Stimulus Properties." *Nature* 338:334–337.

Greenfield, Elliot, and Harold Lecar. 2001. "Mutual Information in a Dilute, Asymmetric Neural Network Model." *Physical Review E* 63:041905.

Griffiths, Robert B. 1969. "Nonanalytic Behavior above the Critical Point in a Random Ising Ferromagnet." *Physical Review Letters* 23:17–19.

Gross, Thilo. 2021. "Not One, But Many Critical States: A Dynamical Systems Perspective." *Frontiers in Neural Circuits* 15:7.

Gu, Mile, Christian Weedbrook, Alvaro Perales, and Michael A. Nielsen. 2009. "More Really Is Different." *Physica D: Nonlinear Phenomena* 238:835–839.

Guitchounts, Grigori. 2020. "An Existential Crisis in Neuroscience." *Nautilus.* New York: Nautilus Think Inc.

Hagemann, Annika, Jens Wilting, Bita Samimizad, Florian Mormann, and Viola Priesemann. 2021. "Assessing Criticality in Pre-Seizure Single-Neuron Activity of Human Epileptic Cortex." *PLOS Computational Biology* 17:e1008773.

Hahn, G., A. Ponce-Alvarez, C. Monier, G. Benvenuti, A. Kumar, F. Chavane, G. Deco, and Y. Fregnac. 2017. "Spontaneous Cortical Activity Is Transiently Poised Close to Criticality." *PLoS Computational Biology* 13:e1005543.

Hahnloser, Richard H. R., Alexay A. Kozhevnikov, and Michale S. Fee. 2002. "An Ultra-Sparse Code Underlies the Generation of Neural Sequences in a Songbird." *Nature* 419:65–70.

Haimovici, Ariel, Enzo Tagliazucchi, Pablo Balenzuela, and Dante R. Chialvo. 2013. "Brain Organization into Resting State Networks Emerges at Criticality on a Model of the Human Connectome." *Physical Review Letters* 110:178101.

Haldeman, C., and J. M. Beggs. 2005. "Critical Branching Captures Activity in Living Neural Networks and Maximizes the Number of Metastable States." *Physical Review Letters* 94:058101.

Hallam, A. 1975. "Alfred Wegener and the Hypothesis of Continental Drift." *Scientific American* 232 (2): 88–97.

Hannun, Awni, Carl Case, Jared Casper, Bryan Catanzaro, Greg Diamos, Erich Elsen, Ryan Prenger, Sanjeev Satheesh, Shubho Sengupta, and Adam Coates. 2014. "Deep Speech: Scaling Up End-to-End Speech Recognition." *arXiv Preprint arXiv:1412.5567.*

Hardstone, Richard, Simon-Shlomo Poil, Giuseppina Schiavone, Rick Jansen, Vadim V. Nikulin, Huibert D. Mansvelder, and Klaus Linkenkaer-Hansen. 2012. "Detrended Fluctuation Analysis: A Scale-Free View on Neuronal Oscillations." *Frontiers in Physiology* 3:450.

Harris, Kenneth D., Jozsef Csicsvari, Hajime Hirase, George Dragoi, and György Buzsáki. 2003. "Organization of Cell Assemblies in the Hippocampus." *Nature* 424:552–556.

Harris, Kenneth D., and Thomas D. Mrsic-Flogel. 2013. "Cortical Connectivity and Sensory Coding." *Nature* 503:51–58.

Harris, Kenneth D., and Gordon M. G. Shepherd. 2015. "The Neocortical Circuit: Themes and Variations." *Nature Neuroscience* 18:170–181.

Hawkins, Jeff, Subutai Ahmad, and Yuwei Cui. 2017. "A Theory of How Columns in the Neocortex Enable Learning the Structure of the World." *Frontiers in Neural Circuits* 11:81.

He, Biyu J., John M. Zempel, Abraham Z. Snyder, and Marcus E. Raichle. 2010. "The Temporal Structures and Functional Significance of Scale-Free Brain Activity." *Neuron* 66:353–369.

He, Kaiming, Xiangyu Zhang, Shaoqing Ren, and Jian Sun. 2016. "Deep Residual Learning for Image Recognition." In *Proceedings of the IEEE Conference on Computer Vision and Pattern Recognition* 770–778.

Heaven, Will Douglas. 2020. "DeepMind's Protein-Folding AI Has Solved a 50-Year-Old Grand Challenge of Biology." *MIT Technology Review*. November 30, 2020.

Hebb, Donald Olding. 2005. *The Organization of Behavior: A Neuropsychological Theory*. Psychology Press.

Helias, Moritz. 2021. "The Brain—as Critical as Possible." *Physics* 14:28.

Hellman, Hal. 2001. *Great Feuds in Medicine: Ten of the Liveliest Disputes Ever*. Hoboken, NJ: John Wiley & Sons.

Hemelrijk, Charlotte K., Lars van Zuidam, and Hanno Hildenbrandt. 2015. "What Underlies Waves of Agitation in Starling Flocks." *Behavioral Ecology and Sociobiology* 69:755–764.

Hengen, Keith B., Mary E. Lambo, Stephen D. Van Hooser, Donald B. Katz, and Gina G. Turrigiano. 2013. "Firing Rate Homeostasis in Visual Cortex of Freely Behaving Rodents." *Neuron* 80:335–342.

Herculano-Houzel, Suzana, Kenneth Catania, Paul R. Manger, and Jon H. Kaas. 2015. "Mammalian Brains Are Made of These: A Dataset of the Numbers and Densities of Neuronal and Nonneuronal Cells in the Brain of Glires, Primates, Scandentia, Eulipotyphlans, Afrotherians and Artiodactyls, and Their Relationship with Body Mass." *Brain, Behavior and Evolution* 86:145–163.

Hernández-Pérez, J. Jesús, Keiland W. Cooper, and Ehren L. Newman. 2020. "Medial Entorhinal Cortex Activates in a Traveling Wave in the Rat." *Elife* 9:e52289.

Herz, Andreas V. M., and John J. Hopfield. 1995. "Earthquake Cycles and Neural Reverberations: Collective Oscillations in Systems with Pulse-Coupled Threshold Elements." *Physical Review Letters* 75:1222.

Herzog, Michael H., and Aaron M. Clarke. 2014. "Why Vision Is Not Both Hierarchical and Feedforward." *Frontiers in Computational Neuroscience* 8:135.

Hilgetag, Claus C., and Marc-Thorsten Hütt. 2014. "Hierarchical Modular Brain Connectivity Is a Stretch for Criticality." *Trends in Cognitive Sciences* 18:114–115.

Hill, Robert Sean, and Christopher A. Walsh. 2005. "Molecular Insights into Human Brain Evolution." *Nature* 437:64–67.

Hobbs, J. P., J. L. Smith, and J. M. Beggs. 2010. "Aberrant Neuronal Avalanches in Cortical Tissue Removed from Juvenile Epilepsy Patients." *Journal of Clinical Neurophysiology* 27:380–386.

Hochstetter, Joel, Ruomin Zhu, Alon Loeffler, Adrian Diaz-Alvarez, Tomonobu Nakayama, and Zdenka Kuncic. 2021. "Avalanches and the Edge-of-Chaos in Neuromorphic Nanowire Networks." *Bulletin of the American Physical Society*.

Hodgkin, Allan L., and Andrew F. Huxley. 1952. "Currents Carried by Sodium and Potassium Ions through the Membrane of the Giant Axon of Loligo." *Journal of Physiology* 116:449–472.

Hofman, Michel A. 2014. "Evolution of the Human Brain: When Bigger Is Better." *Frontiers in Neuroanatomy* 8:15.

Holmgren, Carl, Tibor Harkany, Björn Svennenfors, and Yuri Zilberter. 2003. "Pyramidal Cell Communication within Local Networks in Layer 2/3 of Rat Neocortex." *Journal of Physiology* 551:139–153.

Honey, Christopher J., Thomas Thesen, Tobias H. Donner, Lauren J. Silbert, Chad E. Carlson, Orrin Devinsky, Werner K. Doyle, Nava Rubin, David J. Heeger, and Uri Hasson. 2012. "Slow Cortical Dynamics and the Accumulation of Information over Long Timescales." *Neuron* 76:423–434.

Hopfield, John. J. 1982. "Neural Networks and Physical Systems with Emergent Collective Computational Abilities." *Proceedings of the National Academy of Sciences* 79:2554–2558.

Hopfield, John J. 1984. "Neurons with Graded Response Have Collective Computational Properties Like Those of Two-State Neurons." *Proceedings of the National Academy of Sciences* 81:3088–3092.

Hromádka, Tomáš, Michael R. DeWeese, and Anthony M. Zador. 2008. "Sparse Representation of Sounds in the Unanesthetized Auditory Cortex." *PLoS Biology* 6:e16.

Hsu, D., A. Tang, M. Hsu, and J. M. Beggs. 2007. "Simple Spontaneously Active Hebbian Learning Model: Homeostasis of Activity and Connectivity, and Consequences for Learning and Epileptogenesis." *Physical Review E—Statistical, Nonlinear, and Soft Matter Physics* 76:041909.

Hsu, D., W. Chen, M. Hsu, and J. M. Beggs. 2008. "An Open Hypothesis: Is Epilepsy Learned, and Can It be Unlearned?" *Epilepsy Behavior* 13(3):511–522.

Huang, Xiaoying, William C. Troy, Qian Yang, Hongtao Ma, Carlo R. Laing, Steven J. Schiff, and Jian-Young Wu. 2004. "Spiral Waves in Disinhibited Mammalian Neocortex." *Journal of Neuroscience* 24:9897–9902.

Huang, Xiaoying, Weifeng Xu, Jianmin Liang, Kentaroh Takagaki, Xin Gao, and Jian-young Wu. 2010. "Spiral Wave Dynamics in Neocortex." *Neuron* 68:978–990.

Hutchins, James B., and Steven W. Barger. 1998. "Why Neurons Die: Cell Death in the Nervous System." *The Anatomical Record: An Official Publication of the American Association of Anatomists* 253:79–90.

Hutsler, Jeffrey J., Dong-Geun Lee, and Kristin K. Porter. 2005. "Comparative Analysis of Cortical Layering and Supragranular Layer Enlargement in Rodent Carnivore and Primate Species." *Brain Research* 1052:71–81.

Ihlen, Espen, and Alexander Fürst. 2012. "Introduction to Multifractal Detrended Fluctuation Analysis in Matlab." *Frontiers in Physiology* 3:141.

Inagaki, Hidehiko K., Lorenzo Fontolan, Sandro Romani, and Karel Svoboda. 2019. "Discrete Attractor Dynamics Underlies Persistent Activity in the Frontal Cortex." *Nature* 566:212–217.

Ito, S., F. C. Yeh, E. Hiolski, P. Rydygier, D. E. Gunning, P. Hottowy, N. Timme, A. M. Litke, and J. M. Beggs. 2014. "Large-Scale, High-Resolution Multielectrode-Array Recording Depicts Functional Network Differences of Cortical and Hippocampal Cultures." *PLoS One* 9:e105324.

Iyer, Kartik K., James A. Roberts, Lena Hellström-Westas, Sverre Wikström, Ingrid Hansen Pupp, David Ley, Sampsa Vanhatalo, and Michael Breakspear. 2015. "Cortical Burst Dynamics Predict Clinical Outcome Early in Extremely Preterm Infants." *Brain* 138:2206–2218.

Iyer, Kartik K., James A. Roberts, Marjo Metsäranta, Simon Finnigan, Michael Breakspear, and Sampsa Vanhatalo. 2014. "Novel Features of Early Burst Suppression Predict Outcome after Birth Asphyxia." *Annals of Clinical and Translational Neurology* 1:209–214.

Izhikevich, Eugene M., and Gerald M. Edelman. 2008. "Large-Scale Model of Mammalian Thalamocortical Systems." *Proceedings of the National Academy of Sciences* 105:3593–3598.

Jensen, Henrik Jeldtoft. 1998. *Self-Organized Criticality: Emergent Complex Behavior in Physical and Biological Systems*. Cambridge: Cambridge University Press.

Kaiser, Marcus, Matthias Goerner, and Claus C. Hilgetag. 2007. "Criticality of Spreading Dynamics in Hierarchical Cluster Networks without Inhibition." *New Journal of Physics* 9:110.

Kanda, Paulo Afonso de Medeiros, Renato Anghinah, Magali Taino Smidth, and Jorge Mario Silva. 2009. "The Clinical Use of Quantitative EEG in Cognitive Disorders." *Dementia & Neuropsychologia* 3:195–203.

Kanders, Karlis, Hyungsub Lee, Nari Hong, Yoonkey Nam, and Ruedi Stoop. 2020. "Fingerprints of a Second Order Critical Line in Developing Neural Networks." *Communications Physics* 3:1–13.

Kauffman, Stuart. 1969. "Homeostasis and Differentiation in Random Genetic Control Networks." *Nature* 224:177–178.

Keane, Adam, and Pulin Gong. 2015. "Propagating Waves Can Explain Irregular Neural Dynamics." *Journal of Neuroscience* 35:1591–1605.

Kello, Christopher T. 2013. "Critical Branching Neural Networks." *Psychological Review* 120:230.

Kello, Christopher T., Gordon D. A. Brown, Ramon Ferrer-i-Cancho, John G. Holden, Klaus Linkenkaer-Hansen, Theo Rhodes, and Guy C. Van Orden. 2010. "Scaling Laws in Cognitive Sciences." *Trends in Cognitive Sciences* 14:223–232.

Kelso, J. A. 1984. "Phase Transitions and Critical Behavior in Human Bimanual Coordination." *American Journal of Physiology—Regulatory, Integrative and Comparative Physiology* 246:R1000-R04.

Kelso, J. A. Scott, Guillaume Dumas, and Emmanuelle Tognoli. 2013. "Outline of a General Theory of Behavior and Brain Coordination." *Neural Networks* 37:120–131.

Kelso, Stephen R., Alan H. Ganong, and Thomas H. Brown. 1986. "Hebbian Synapses in Hippocampus." *Proceedings of the National Academy of Sciences* 83:5326–5330.

Kilpatrick, Zachary P. 2014. "Wilson-Cowan Model." In *Encyclopedia of Computational Neuroscience*. New York: Springer.

Kim, Dal Hyung, Jungsoo Kim, João C. Marques, Abhinav Grama, David G. C. Hildebrand, Wenchao Gu, Jennifer M. Li, and Drew N. Robson. 2017. "Pan-Neuronal Calcium Imaging with Cellular Resolution in Freely Swimming Zebrafish." *Nature Methods* 14:1107–1114.

Kinouchi, Osame, and Mauro Copelli. 2006. "Optimal Dynamical Range of Excitable Networks at Criticality." *Nature Physics*, 2:348–51.

Kitzbichler, Manfred G., Marie L. Smith, Søren R. Christensen, and Ed Bullmore. 2009. "Broadband Criticality of Human Brain Network Synchronization." *PLoS Computational Biology* 5:e1000314.

Kostinski, S. and A. Amir, 2016. "An Elementary Derivation of First and Last Return Times of 1D Random Walks." *American Journal of Physics*, 84:57–60.

Krizhevsky, Alex, Ilya Sutskever, and Geoffrey E. Hinton. 2012. "Imagenet Classification with Deep Convolutional Neural Networks." *Advances in Neural Information Processing Systems* 25:1097–1105.

Kuramoto, Y. 1984. *Chemical Oscillations, Waves, and Turbulence*. Berlin: Springer Verlag.

Lee, Hyodong, Eshed Margalit, Kamila M. Jozwik, Michael A. Cohen, Nancy Kanwisher, Daniel L. K. Yamins, and James J. DiCarlo. 2020. "Topographic Deep Artificial Neural Networks Reproduce the Hallmarks of the Primate Inferior Temporal Cortex Face Processing Network." *bioRxiv*.

Lefort, Sandrine, Christian Tomm, J.-C. Floyd Sarria, and Carl C. H. Petersen. 2009. "The Excitatory Neuronal Network of the C2 Barrel Column in Mouse Primary Somatosensory Cortex." *Neuron* 61:301–316.

Lennie, Peter. 1998. "Single Units and Visual Cortical Organization." *Perception* 27:889–935.

Levina, Anna, J. Michael Herrmann, and Theo Geisel. 2007. "Dynamical Synapses Causing Self-Organized Criticality in Neural Networks." *Nature Physics* 3:857–860.

Levina, Anna, and Viola Priesemann. 2017. "Subsampling Scaling." *Nature Communications* 8:1–9.

Lin, Henry W., Max Tegmark, and David Rolnick. 2017. "Why Does Deep and Cheap Learning Work So Well?" *Journal of Statistical Physics* 168:1223–1247.

Linkenkaer-Hansen, Klaus, Vadim V. Nikouline, J. Matias Palva, and Risto J. Ilmoniemi. 2001. "Long-Range Temporal Correlations and Scaling Behavior in Human Brain Oscillations." *Journal of Neuroscience* 21:1370–1377.

Lombardi, Fabrizio, Hans J. Herrmann, Carla Perrone-Capano, Dietmar Plenz, and Lucilla De Arcangelis. 2012. "Balance between Excitation and Inhibition Controls the Temporal Organization of Neuronal Avalanches." *Physical Review Letters* 108:228703.

Lombardi, Fabrizio, Hans J. Herrmann, Dietmar Plenz, and Lucilla De Arcangelis. 2016. "Temporal Correlations in Neuronal Avalanche Occurrence." *Scientific Reports* 6:1–12.

Louie, Kenway, and Matthew A. Wilson. 2001. "Temporally Structured Replay of Awake Hippocampal Ensemble Activity during Rapid Eye Movement Sleep." *Neuron* 29:145–156.

Lubenov, Evgueniy V., and Athanassios G. Siapas. 2009. "Hippocampal Theta Oscillations Are Travelling Waves." *Nature* 459:534–539.

Ma, Zhengyu, Haixin Liu, Takaki Komiyama, and Ralf Wessel. 2020. "Stability of Motor Cortex Network States during Learning-Associated Neural Reorganizations." *Journal of Neurophysiology* 124:1327–1342.

Ma, Zhengyu, Gina G. Turrigiano, Ralf Wessel, and Keith B. Hengen. 2019. "Cortical Circuit Dynamics Are Homeostatically Tuned to Criticality In Vivo." *Neuron* 104:655–664, e4.

Maass, Wolfgang, Thomas Natschläger, and Henry Markram. 2002. "Real-Time Computing without Stable States: A New Framework for Neural Computation Based on Perturbations." *Neural Computation* 14:2531–2560.

Magnasco, Marcelo O. 2003. "A Wave Traveling over a Hopf Instability Shapes the Cochlear Tuning Curve." *Physical Review Letters* 90:058101.

Magri, Cesare, Kevin Whittingstall, Vanessa Singh, Nikos K. Logothetis, and Stefano Panzeri. 2009. "A Toolbox for the Fast Information Analysis of Multiple-Site LFP, EEG and Spike Train Recordings." *BMC Neuroscience* 10:81.

Malow, Beth A. 2004. "Sleep Deprivation and Epilepsy." *Epilepsy Currents* 4:193–195.

Mariani, Benedetta, Giorgio Nicoletti, Marta Bisio, Marta Maschietto, Roberto Oboe, Samir Suweis, and Stefano Vassanelli. 2021. "Beyond Resting State Neuronal Avalanches in the Somatosensory Barrel Cortex." *bioRxiv.*

Marinazzo, D., M. Pellicoro, G. Wu, L. Angelini, J. M. Cortés, and S. Stramaglia. 2014. "Information Transfer and Criticality in the Ising Model on the Human Connectome." *PloS One* 9(4).

Mariño, Jorge, James Schummers, David C. Lyon, Lars Schwabe, Oliver Beck, Peter Wiesing, Klaus Obermayer, and Mriganka Sur. 2005. "Invariant Computations in Local Cortical Networks with Balanced Excitation and Inhibition." *Nature Neuroscience* 8:194–201.

Marković, Dimitrije, and Claudius Gros. 2014. "Power Laws and Self-Organized Criticality in Theory and Nature." *Physics Reports* 536:41–74.

Markram, Henry. 2006. "The Blue Brain Project." *Nature Reviews Neuroscience* 7:153–160.

Markram, Henry, Eilif Muller, Srikanth Ramaswamy, Michael W. Reimann, Marwan Abdellah, Carlos Aguado Sanchez, Anastasia Ailamaki, Lidia Alonso-Nanclares, Nicolas Antille, and Selim Arsever. 2015. "Reconstruction and Simulation of Neocortical Microcircuitry." *Cell* 163:456–492.

Marshall, N., N. M. Timme, N. Bennett, M. Ripp, E. Lautzenhiser, and J. M. Beggs. 2016. "Analysis of Power Laws, Shape Collapses, and Neural Complexity: New Techniques and MATLAB Support via the NCC Toolbox." *Frontiers in Physiology* 7:250.

Martinello, Matteo, Jorge Hidalgo, Amos Maritan, Serena Di Santo, Dietmar Plenz, and Miguel A. Muñoz. 2017. "Neutral Theory and Scale-Free Neural Dynamics." *Physical Review X* 7:041071.

Mason, Adrian, Andrew Nicoll, and Ken Stratford. 1991. "Synaptic Transmission between Individual Pyramidal Neurons of the Rat Visual Cortex In Vitro." *Journal of Neuroscience* 11:72–84.

Meisel, Christian. 2020. "Antiepileptic Drugs Induce Subcritical Dynamics in Human Cortical Networks." *Proceedings of the National Academy of Sciences* 117:11118–11125.

Meisel, Christian, Kimberlyn Bailey, Peter Achermann, and Dietmar Plenz. 2017. "Decline of Long-Range Temporal Correlations in the Human Brain during Sustained Wakefulness." *Scientific Reports* 7:1–11.

Meisel, Christian, Eckehard Olbrich, Oren Shriki, and Peter Achermann. 2013. "Fading Signatures of Critical Brain Dynamics during Sustained Wakefulness in Humans." *Journal of Neuroscience* 33:17363–17372.

Meisel, Christian, Andreas Schulze-Bonhage, Dean Freestone, Mark James Cook, Peter Achermann, and Dietmar Plenz. 2015. "Intrinsic Excitability Measures Track Antiepileptic Drug Action and Uncover Increasing/

Decreasing Excitability over the Wake/Sleep Cycle." *Proceedings of the National Academy of Sciences* 112: 14694–14699.

Meisel, Christian, Alexander Storch, Susanne Hallmeyer-Elgner, Ed Bullmore, and Thilo Gross. 2012. "Failure of Adaptive Self-Organized Criticality during Epileptic Seizure Attacks." *PLoS Computational Biology* 8:e1002312.

Meshulam, Leenoy, Jeffrey L. Gauthier, Carlos D. Brody, David W. Tank, and William Bialek. 2019. "Coarse Graining, Fixed Points, and Scaling in a Large Population of Neurons." *Physical Review Letters* 123:178103.

Mikkelsen, Tarjei S., LaDeana W. Hillier, Evan E. Eichler, Michael C. Zody, David B. Jaffe, Shiaw-Pyng Yang, Wolfgang Enard, Ines Hellmann, Kerstin Lindblad-Toh, and Tasha K. Altheide. 2005. "Initial Sequence of the Chimpanzee Genome and Comparison with the Human Genome." *Nature* 437:69–87.

Miller, Stephanie R., Shan Yu, and Dietmar Plenz. 2019. "The Scale-Invariant, Temporal Profile of Neuronal Avalanches in Relation to Cortical γ–Oscillations." *Scientific Reports* 9:1–14.

Millman, Daniel, Stefan Mihalas, Alfredo Kirkwood, and Ernst Niebur. 2010. "Self-Organized Criticality Occurs in Non-Conservative Neuronal Networks during 'Up' States." *Nature Physics* 6:801–805.

Mitzenmacher, M. 2004. "A Brief History of Generative Models for Power Law and Lognormal Distributions." *Internet Mathematics* 1 (2): 226–251.

Molchanov, Pavlo, Stephen Tyree, Tero Karras, Timo Aila, and Jan Kautz. 2016. "Pruning Convolutional Neural Networks for Resource Efficient Inference." *arXiv Preprint arXiv:1611.06440*.

Montez, Teresa, Simon-Shlomo Poil, Bethany F. Jones, Ilonka Manshanden, Jeroen P. A. Verbunt, Bob W. van Dijk, Arjen B. Brussaard, Arjen van Ooyen, Cornelis J. Stam, and Philip Scheltens. 2009. "Altered Temporal Correlations in Parietal Alpha and Prefrontal Theta Oscillations in Early-Stage Alzheimer Disease." *Proceedings of the National Academy of Sciences* 106:1614–1619.

Mora, Thierry, and William Bialek. 2011. "Are Biological Systems Poised at Criticality?" *Journal of Statistical Physics* 144:268–302.

Mora, Thierry, Aleksandra M. Walczak, William Bialek, and Curtis G. Callan. 2010. "Maximum Entropy Models for Antibody Diversity." *Proceedings of the National Academy of Sciences* 107:5405–5410.

Moretti, Paolo, and Miguel A. Muñoz. 2013. "Griffiths Phases and the Stretching of Criticality in Brain Networks." *Nature Communications* 4:1–10.

Muller, Lyle, Frédéric Chavane, John Reynolds, and Terrence J. Sejnowski. 2018. "Cortical Travelling Waves: Mechanisms and Computational Principles." *Nature Reviews Neuroscience* 19:255.

Muller, Lyle, Alexandre Reynaud, Frédéric Chavane, and Alain Destexhe. 2014. "The Stimulus-Evoked Population Response in Visual Cortex of Awake Monkey is a Propagating Wave." *Nature Communications* 5:1–14.

Muñoz, Miguel A., Róbert Juhász, Claudio Castellano, and Géza Odor. 2010. "Griffiths Phases on Complex Networks." *Physical Review Letters* 105:128701.

Murray, John D., Alberto Bernacchia, David J. Freedman, Ranulfo Romo, Jonathan D. Wallis, Xinying Cai, Camillo Padoa-Schioppa, Tatiana Pasternak, Hyojung Seo, and Daeyeol Lee. 2014. "A Hierarchy of Intrinsic Timescales across Primate Cortex." *Nature Neuroscience* 17:1661–1663.

Nematzadeh, Azadeh, Emilio Ferrara, Alessandro Flammini, and Yong-Yeol Ahn. 2014. "Optimal Network Modularity for Information Diffusion." *Physical Review Letters* 113:088701.

Netoff, Theoden I., Robert Clewley, Scott Arno, Tara Keck, and John A. White. 2004. "Epilepsy in Small-World Networks." *Journal of Neuroscience* 24:8075–8083.

Nigam, S., M. Shimono, S. Ito, F. C. Yeh, N. Timme, M. Myroshnychenko, C. C. Lapish, Z. Tosi, P. Hottowy, W. C. Smith, S. C. Masmanidis, A. M. Litke, O. Sporns, and J. M. Beggs. 2016. "Rich-Club Organization in Effective Connectivity among Cortical Neurons." *Journal of Neuroscience* 36:670–684.

Nishimori, Hidetoshi, and Gerardo Ortiz. 2010. *Elements of Phase Transitions and Critical Phenomena.* Oxford: Oxford University Press.

Niss, Martin. 2005. "History of the Lenz-Ising Model 1920–1950: From Ferromagnetic to Cooperative Phenomena." *Archive for History of Exact Sciences* 59:267–318.

Nordfalk, Jacob, and Preben Alstrøm. 1996. "Phase Transitions near the 'Game of Life.'" *Physical Review E* 54:R1025.

O'Connor, Timothy. 1994. "Emergent Properties." *American Philosophical Quarterly* 31:91–104.

O'Donohue, Thomas L., William R. Millington, Gail E. Handelmann, Patricia C. Contreras, and Bibie M. Chronwall. 1985. "On the 50th Anniversary of Dale's Law: Multiple Neurotransmitter Neurons." *Trends in Pharmacological Sciences* 6:305–308.

O'Reilly, Randall C., and Yuko Munakata. 2000. *Computational Explorations in Cognitive Neuroscience: Understanding the Mind by Simulating the Brain.* Cambridge, MA: MIT Press.

Ottman, Ruth, John F. Annegers, Neil Risch, W. Allen Hauser, and Mervyn Susser. 1996. "Relations of Genetic and Environmental Factors in the Etiology of Epilepsy." *Annals of Neurology* 39:442–449.

Ouellette, Jennifer. 2018. "Brains May Teeter near Their Tipping Point." *Quanta Magazine*. Simons Foundation.

Pajevic, Sinisa, and Dietmar Plenz. 2009. "Efficient Network Reconstruction from Dynamical Cascades Identifies Small-World Topology of Neuronal Avalanches." *PLoS Computational Biology* 5:e1000271.

Palomero-Gallagher, Nicola, and Karl Zilles. 2019. "Cortical Layers: Cyto-, Myelo-, Receptor-and Synaptic Architecture in Human Cortical Areas." *Neuroimage* 197:716–741.

Palva, J. Matias, Alexander Zhigalov, Jonni Hirvonen, Onerva Korhonen, Klaus Linkenkaer-Hansen, and Satu Palva. 2013. "Neuronal Long-Range Temporal Correlations and Avalanche Dynamics Are Correlated with Behavioral Scaling Laws." *Proceedings of the National Academy of Sciences* 110:3585–3590.

Panas, Dagmara, Hayder Amin, Alessandro Maccione, Oliver Muthmann, Mark van Rossum, Luca Berdondini, and Matthias H. Hennig. 2015. "Sloppiness in Spontaneously Active Neuronal Networks." *Journal of Neuroscience* 35:8480–8492.

Panzeri, Stefano, Riccardo Senatore, Marcelo A. Montemurro, and Rasmus S. Petersen. 2007. "Correcting for the Sampling Bias Problem in Spike Train Information Measures." *Journal of Neurophysiology* 98:1064–1072.

Pasquale, V., P. Massobrio, L. L. Bologna, M. Chiappalone, and S. Martinoia. 2008. "Self-Organization and Neuronal Avalanches in Networks of Dissociated Cortical Neurons." *Neuroscience* 153:1354–1369.

Peng, Jiayi, and John M. Beggs. 2013. "Attaining and Maintaining Criticality in a Neuronal Network Model." *Physica A: Statistical Mechanics and Its Applications* 392:1611–1620.

Perin, Rodrigo, Thomas K. Berger, and Henry Markram. 2011. "A Synaptic Organizing Principle for Cortical Neuronal Groups." *Proceedings of the National Academy of Sciences* 108:5419–5424.

Petermann, Thomas, Tara C. Thiagarajan, Mikhail A. Lebedev, Miguel A. L. Nicolelis, Dante R. Chialvo, and Dietmar Plenz. 2009. "Spontaneous Cortical Activity in Awake Monkeys Composed of Neuronal Avalanches." *Proceedings of the National Academy of Sciences* 106:15921–15926.

Peters, Andrew J., Simon X. Chen, and Takaki Komiyama. 2014. "Emergence of Reproducible Spatiotemporal Activity during Motor Learning." *Nature* 510:263–267.

Peters, Andrew J., Jun Lee, Nathan G. Hedrick, Keelin O'Neil, and Takaki Komiyama. 2017. "Reorganization of Corticospinal Output during Motor Learning." *Nature Neuroscience* 20:1133.

Pike, Matthew D., Saurabh K. Bose, Joshua B. Mallinson, Susant K. Acharya, Shota Shirai, Edoardo Galli, Stephen J. Weddell, Philip J. Bones, Matthew D. Arnold, and Simon A. Brown. 2020. "Atomic Scale Dynamics Drive Brain-Like Avalanches in Percolating Nanostructured Networks." *Nano Letters* 20:3935–3942.

Pipkin, Jason. 2020. "Connectomes: Mapping the Mind of a Fly." *Elife* 9:e62451.

Plenz, Dietmar, and Ernst Niebur. 2014. *Criticality in Neural Systems*. Hoboken, NJ: John Wiley & Sons.

Poggio, Tomaso. 1990. "A Theory of How the Brain Might Work." In *Cold Spring Harbor Symposia on Quantitative Biology*, 899–910. Cold Spring Harbor Laboratory Press.

Poil, Simon-Shlomo, Richard Hardstone, Huibert D. Mansvelder, and Klaus Linkenkaer-Hansen. 2012. "Critical-State Dynamics of Avalanches and Oscillations Jointly Emerge from Balanced Excitation/Inhibition in Neuronal Networks." *Journal of Neuroscience* 32:9817–9823.

Ponce-Alvarez, A., A. Jouary, M. Privat, G. Deco, and G. Sumbre. 2018. "Whole-Brain Neuronal Activity Displays Crackling Noise Dynamics." *Neuron* 100 (6): 1446–1459 e1446.

Priesemann, Viola, Mario Valderrama, Michael Wibral, and Michel Le Van Quyen. 2013. "Neuronal Avalanches Differ from Wakefulness to Deep Sleep—Evidence from Intracranial Depth Recordings in Humans." *PLoS Computational Biology* 9 (3).

Priesemann, Viola, Michael Wibral, Mario Valderrama, Robert Pröpper, Michel Le Van Quyen, Theo Geisel, Jochen Triesch, Danko Nikolić, and Matthias H. J. Munk. 2014. "Spike Avalanches In Vivo Suggest a Driven, Slightly Subcritical Brain State." *Frontiers in Systems Neuroscience* 8:108.

Proekt, Alex, Jayanth R. Banavar, Amos Maritan, and Donald W. Pfaff. 2012. "Scale Invariance in the Dynamics of Spontaneous Behavior." *Proceedings of the National Academy of Sciences* 109:10564–10569.

Prosi, Jan, Sina Khajehabdollahi, Emmanouil Giannakakis, Georg Martius, and Anna Levina. 2021. "The Dynamical Regime and Its Importance for Evolvability, Task Performance and Generalization." *arXiv Preprint arXiv:2103.12184.*

Radonjić, Ana, Sarah R. Allred, Alan L. Gilchrist, and David H. Brainard. 2011. "The Dynamic Range of Human Lightness Perception." *Current Biology* 21:1931–1936.

Rajan, Kanaka, and Larry F. Abbott. 2006. "Eigenvalue Spectra of Random Matrices for Neural Networks." *Physical Review Letters* 97:188104.

Rakic, Pasko. 2009. "Evolution of the Neocortex: A Perspective from Developmental Biology." *Nature Reviews Neuroscience* 10:724–735.

Ramirez, Jonatan Pena, and Henk Nijmeijer. 2020. "The Secret of the Synchronized Pendulums." *Physics World* 33:36.

Ramirez, Steve, Xu Liu, Pei-Ann Lin, Junghyup Suh, Michele Pignatelli, Roger L. Redondo, Tomás J. Ryan, and Susumu Tonegawa. 2013. "Creating a False Memory in the Hippocampus." *Science* 341:387–391.

Reed, W. J., and B. D. Hughes. 2002. "From Gene Families and Genera to Incomes and Internet File Sizes: Why Power Laws Are So Common in Nature." *Physical Review E* 66 (6): 067103.

Rendell, Paul. 2011. "A Universal Turing Machine in Conway's Game of Life." In *2011 International Conference on High Performance Computing & Simulation*, 764–772. IEEE.

Roberts, James A., Leonardo L. Gollo, Romesh G. Abeysuriya, Gloria Roberts, Philip B. Mitchell, Mark W. Woolrich, and Michael Breakspear. 2019. "Metastable Brain Waves." *Nature Communications* 10:1–17.

Roberts, James A., Kartik K. Iyer, Simon Finnigan, Sampsa Vanhatalo, and Michael Breakspear. 2014. "Scale-Free Bursting in Human Cortex Following Hypoxia at Birth." *Journal of Neuroscience* 34:6557–6572.

Rolnick, David, and Max Tegmark. 2017. "The Power of Deeper Networks for Expressing Natural Functions." *arXiv Preprint arXiv:1705.05502*.

Rolston, John D., Daniel A. Wagenaar, and Steve M. Potter. 2007. "Precisely Timed Spatiotemporal Patterns of Neural Activity in Dissociated Cortical Cultures." *Neuroscience* 148:294–303.

Rosenblatt, Frank. 1958. "The Perceptron: A Probabilistic Model for Information Storage and Organization in the Brain." *Psychological Review* 65:386.

Roser, M., C. Appel, and H. Ritchie. 2013. "Human Height." Published online at OurWorldInData.org, https://ourworldindata.org/human-height.

Rubino, Doug, Kay A. Robbins, and Nicholas G. Hatsopoulos. 2006. "Propagating Waves Mediate Information Transfer in the Motor Cortex." *Nature Neuroscience* 9:1549–1557.

Rubinov, Mikail, Olaf Sporns, Jean-Philippe Thivierge, and Michael Breakspear. 2011. "Neurobiologically Realistic Determinants of Self-Organized Criticality in Networks of Spiking Neurons." *PLoS Computational Biology* 7:e1002038.

Ruderman, Daniel L., and William Bialek. 1994. "Statistics of Natural Images: Scaling in the Woods." *Physical Review Letters* 73:814.

Saha, Debajit, David Morton, Michael Ariel, and Ralf Wessel. 2011. "Response Properties of Visual Neurons in the Turtle Nucleus Isthmi." *Journal of Comparative Physiology A* 197:153–165.

Salzman, C. Daniel, Kenneth H. Britten, and William T. Newsome. 1990. "Cortical Microstimulation Influences Perceptual Judgements of Motion Direction." *Nature* 346:174–177.

Sanchez-Vives, Maria V., and David A. McCormick. 2000. "Cellular and Network Mechanisms of Rhythmic Recurrent Activity in Neocortex." *Nature Neuroscience* 3:1027–1034.

Saremi, Saeed, and Terrence J. Sejnowski. 2013. "Hierarchical Model of Natural Images and the Origin of Scale Invariance." *Proceedings of the National Academy of Sciences* 110:3071–3076.

Scarpetta, Silvia, Ilenia Apicella, Ludovico Minati, and Antonio de Candia. 2018. "Hysteresis, Neural Avalanches, and Critical Behavior near a First-Order Transition of a Spiking Neural Network." *Physical Review E* 97:062305.

Schiff, Steven J., Kristin Jerger, Duc H. Duong, Taeun Chang, Mark L. Spano, and William L. Ditto. 1994. "Controlling Chaos in the Brain." *Nature* 370:615–620.

Schneidman, Elad, Michael J. Berry, Ronen Segev, and William Bialek. 2006. "Weak Pairwise Correlations Imply Strongly Correlated Network States in a Neural Population." *Nature* 440:1007–1012.

Schoenemann, P. Thomas. 2006. "Evolution of the Size and Functional Areas of the Human Brain." *Annual Review of Anthropology*, 35:379–406.

Scott, Gregory, Erik D. Fagerholm, Hiroki Mutoh, Robert Leech, David J. Sharp, Woodrow L. Shew, and Thomas Knöpfel. 2014. "Voltage Imaging of Waking Mouse Cortex Reveals Emergence of Critical Neuronal Dynamics." *Journal of Neuroscience* 34:16611–16620.

Sethna, James P., Karin A. Dahmen, and Christopher R. Myers. 2001. "Crackling Noise." *Nature* 410:242–250.

Sharma, Jitendra, Alessandra Angelucci, and Mriganka Sur. 2000. "Induction of Visual Orientation Modules in Auditory Cortex." *Nature* 404:841–847.

Sharp, Kim, and Franz Matschinsky. 2015. "Translation of Ludwig Boltzmann's Paper 'On the Relationship between the Second Fundamental Theorem of the Mechanical Theory of Heat and Probability Calculations Regarding the Conditions for Thermal Equilibrium,' Sitzungberichte der Kaiserlichen Akademie der Wissenschaften. Mathematisch-Naturwissen Classe. Abt. II, LXXVI 1877, pp. 373–435 (Wien. Ber. 1877, 76: 373–435). Reprinted in Wiss. Abhandlungen, Vol. II, reprint 42, pp. 164–223, Barth, Leipzig, 1909." *Entropy* 17:1971–2009.

Shaw, Philip, Deanna Greenstein, Jason Lerch, Liv Clasen, Rhoshel Lenroot, N. E. E. A. Gogtay, Alan Evans, J. Rapoport, and J. Giedd. 2006. "Intellectual Ability and Cortical Development in Children and Adolescents." *Nature* 440:676–679.

Shaw, Philip, Noor J. Kabani, Jason P. Lerch, Kristen Eckstrand, Rhoshel Lenroot, Nitin Gogtay, Deanna Greenstein, Liv Clasen, Alan Evans, and Judith L. Rapoport. 2008. "Neurodevelopmental Trajectories of the Human Cerebral Cortex." *Journal of Neuroscience* 28:3586–3594.

Shepherd, Gordon M. 2004. *The Synaptic Organization of the Brain.* Oxford: Oxford University Press.

Sherwood, S. 2011. "Science Controversies Past and Present." *Physics Today* 64 (10): 39.

Shew, W. L., W. P. Clawson, J. Pobst, Y. Karimipanah, N. C. Wright and R. Wessel. 2015. "Adaptation to Sensory Input Tunes Visual Cortex to Criticality." *Nature Physics* 11 (8): 659.

Shew, W. L., and D. Plenz. 2013. "The Functional Benefits of Criticality in the Cortex." *Neuroscientist* 19:88–100.

Shew, W. L., H. Yang, T. Petermann, R. Roy, and D. Plenz. 2009. "Neuronal Avalanches Imply Maximum Dynamic Range in Cortical Networks at Criticality." *Journal of Neuroscience* 29:15595–15600.

Shew, W. L., H. Yang, S. Yu, R. Roy, and D. Plenz. 2011. "Information Capacity and Transmission Are Maximized in Balanced Cortical Networks with Neuronal Avalanches." *Journal of Neuroscience* 31:55–63.

Shimono, M., and J. M. Beggs. 2015. "Functional Clusters, Hubs, and Communities in the Cortical Microconnectome." *Cerebral Cortex* 25:3743–3757.

Shipp, Stewart. 2005. "The Importance of Being Agranular: A Comparative Account of Visual and Motor Cortex." *Philosophical Transactions of the Royal Society B: Biological Sciences* 360:797–814.

Shirai, Shota, Susant Kumar Acharya, Saurabh Kumar Bose, Joshua Brian Mallinson, Edoardo Galli, Matthew D. Pike, Matthew D. Arnold, and Simon Anthony Brown. 2020. "Long-Range Temporal Correlations in Scale-Free Neuromorphic Networks." *Network Neuroscience* 4:432–447.

Shmulevich, Ilya, Stuart A. Kauffman, and Maximino Aldana. 2005. "Eukaryotic Cells Are Dynamically Ordered or Critical but not Chaotic." *Proceedings of the National Academy of Sciences* 102:13439–13444.

Shriki, Oren, Jeff Alstott, Frederick Carver, Tom Holroyd, Richard N. A. Henson, Marie L. Smith, Richard Coppola, Edward Bullmore, and Dietmar Plenz. 2013. "Neuronal Avalanches in the Resting MEG of the Human Brain." *Journal of Neuroscience* 33:7079–7090.

Siegle, Joshua H., Xiaoxuan Jia, Séverine Durand, Sam Gale, Corbett Bennett, Nile Graddis, Greggory Heller, Tamina K. Ramirez, Hannah Choi, and Jennifer A. Luviano. 2021. "Survey of Spiking in the Mouse Visual System Reveals Functional Hierarchy." *Nature* 592:86–92.

Silver, David, Aja Huang, Chris J. Maddison, Arthur Guez, Laurent Sifre, George Van Den Driessche, Julian Schrittwieser, Ioannis Antonoglou, Veda Panneershelvam, and Marc Lanctot. 2016. "Mastering the Game of Go with Deep Neural Networks and Tree Search." *Nature* 529:484–489.

Singer, Wolf. 1999. "Neuronal Synchrony: A Versatile Code for the Definition of Relations?" *Neuron* 24:49–65.

Sinha, Sudeshna, and William L. Ditto. 1999. "Computing with Distributed Chaos." *Physical Review E* 60:363.

Smith, Jeffrey C., Howard H. Ellenberger, Klaus Ballanyi, Diethelm W. Richter, and Jack L. Feldman. 1991. "Pre-Bötzinger Complex: A Brainstem Region That May Generate Respiratory Rhythm in Mammals." *Science* 254:726–729.

Smith, Shelagh J. M. 2005. "EEG in the Diagnosis, Classification, and Management of Patients with Epilepsy." *Journal of Neurology, Neurosurgery & Psychiatry* 76:ii2–ii7.

Solovey, Guillermo, Leandro M. Alonso, Toru Yanagawa, Naotaka Fujii, Marcelo O. Magnasco, Guillermo A. Cecchi, and Alex Proekt. 2015. "Loss of Consciousness Is Associated with Stabilization of Cortical Activity." *Journal of Neuroscience* 35:10866–10877.

Song, Sen, and Larry F. Abbott. 2001. "Cortical Development and Remapping through Spike Timing-Dependent Plasticity." *Neuron* 32:339–350.

Song, Sen, Per Jesper Sjöström, Markus Reigl, Sacha Nelson, and Dmitri B. Chklovskii. 2005. "Highly Non-random Features of Synaptic Connectivity in Local Cortical Circuits." *PLoS Biology* 3.

Spitzner, F. P., J. Dehning, J. Wilting, A. Hagemann, J. P. Neto, J. Zierenberg, and V. Priesemann. 2020. "MR. Estimator, a Toolbox to Determine Intrinsic Timescales from Subsampled Spiking Activity." *arXiv Preprint arXiv:2007.03367.*

Stanley, H. E. 1971. *Introduction to Phase Transitions and Critical Phenomena.* Oxford: Clarendon Press.

Stepp, Nigel, Dietmar Plenz, and Narayan Srinivasa. 2015. "Synaptic Plasticity Enables Adaptive Self-Tuning Critical Networks." *PLOS Computational Biology* 11:e1004043.

Stewart, Craig V., and Dietmar Plenz. 2006. "Inverted-U Profile of Dopamine–NMDA-Mediated Spontaneous Avalanche Recurrence in Superficial Layers of Rat Prefrontal Cortex." *Journal of Neuroscience* 26:8148–8159.

Stewart, Craig V., and Dietmar Plenz. 2008. "Homeostasis of Neuronal Avalanches During Postnatal Cortex Development In Vitro." *Journal of Neuroscience Methods* 169:405–416.

Stieg, A. Z., A. V. Avizienis, H. O. Sillin, C. Martin-Olmos, M. Aono, and J. K. Gimzewski. 2012. "Emergent Criticality in Complex Turing B-Type Atomic Switch Networks." *Advanced Materials* 24 (2): 286–293.

Stringer, Carsen, Marius Pachitariu, Nicholas Steinmetz, Matteo Carandini, and Kenneth D. Harris. 2019. "High-Dimensional Geometry of Population Responses in Visual Cortex." *Nature* 571:361–365.

Sun, Wenzhi, and Yang Dan. 2009. "Layer-Specific Network Oscillation and Spatiotemporal Receptive Field in the Visual Cortex." *Proceedings of the National Academy of Sciences* 106:17986–17991.

Tagliazucchi, Enzo, Pablo Balenzuela, Daniel Fraiman, and Dante R. Chialvo. 2012. "Criticality in Large-Scale Brain fMRI Dynamics Unveiled by a Novel Point Process Analysis." *Frontiers in Physiology* 3:15.

Tagliazucchi, Enzo, Dante R. Chialvo, Michael Siniatchkin, Enrico Amico, Jean-Francois Brichant, Vincent Bonhomme, Quentin Noirhomme, Helmut Laufs, and Steven Laureys. 2016. "Large-Scale Signatures of Unconsciousness Are Consistent with a Departure from Critical Dynamics." *Journal of the Royal Society Interface* 13:20151027.

Teets, Donald, and Karen Whitehead. 1999. "The Discovery of Ceres: How Gauss Became Famous." *Mathematics Magazine* 72:83–93.

Teich, Malvin C., Conor Heneghan, Steven B. Lowen, Tsuyoshi Ozaki, and Ehud Kaplan. 1997. "Fractal Character of the Neural Spike Train in the Visual System of the Cat." *JOSA A* 14:529–546.

Tetzlaff, C., S. Okujeni, U. Egert, F. Worgotter, and M. Butz. 2010. "Self-Organized Criticality in Developing Neuronal Networks." *PLoS Computational Biology* 6:e1001013.

Timme, Nicholas M., Najja J. Marshall, Nicholas Bennett, Monica Ripp, Edward Lautzenhiser, and John M. Beggs. 2016. "Criticality Maximizes Complexity in Neural Tissue." *Frontiers in Physiology* 7:425.

Tkačik, Gašper, Thierry Mora, Olivier Marre, Dario Amodei, Stephanie E. Palmer, Michael J. Berry, and William Bialek. 2015. "Thermodynamics and Signatures of Criticality in a Network of Neurons." *Proceedings of the National Academy of Sciences* 112:11508–11513.

Tkačik, Gašper, Elad Schneidman, Michael J. Berry II, and William Bialek. 2009. "Spin Glass Models for a Network of Real Neurons." *arXiv Preprint arXiv:0912.5409.*

Toet, M. C., Lena Hellström-Westas, F. Groenendaal, P. Eken, and L. S. De Vries. 1999. "Amplitude Integrated EEG 3 and 6 Hours After Birth in Full Term Neonates with Hypoxic–Ischaemic Encephalopathy." *Archives of Disease in Childhood—Fetal and Neonatal Edition* 81:F19–F23.

Tomen, Nergis, J. Michael Herrmann, and Udo Ernst. 2019. *The Functional Role of Critical Dynamics in Neural Systems.* New York: Springer.

Tononi, Giulio, and Gerald M. Edelman. 1998. "Consciousness and Complexity." *Science* 282:1846–1851.

Tononi, Giulio, Olaf Sporns, and Gerald M. Edelman. 1994. "A Measure for Brain Complexity: Relating Functional Segregation and Integration in the Nervous System." *Proceedings of the National Academy of Sciences* 91:5033–5037.

Touboul, Jonathan, and Alain Destexhe. 2010. "Can Power-Law Scaling and Neuronal Avalanches Arise from Stochastic Dynamics?" *PLoS One* 5 (2): e8982.

Touboul, Jonathan, and Alain Destexhe. 2017. "Power-Law Statistics and Universal Scaling in the Absence of Criticality." *Physical Review E* 95:012413.

Turrigiano, Gina G., Kenneth R. Leslie, Niraj S. Desai, Lana C. Rutherford, and Sacha B. Nelson. 1998. "Activity-Dependent Scaling of Quantal Amplitude in Neocortical Neurons." *Nature* 391:892–896.

Turrigiano, Gina G., and Sacha B. Nelson. 2004. "Homeostatic Plasticity in the Developing Nervous System." *Nature Reviews Neuroscience* 5:97–107.

Van Vreeswijk, Carl, and Haim Sompolinsky. 1996. "Chaos in Neuronal Networks with Balanced Excitatory and Inhibitory Activity." *Science* 274:1724–1726.

Varela, Juan A., Kamal Sen, Jay Gibson, Joshua Fost, L. F. Abbott, and Sacha B. Nelson. 1997. "A Quantitative Description of Short-Term Plasticity at Excitatory Synapses in Layer 2/3 of Rat Primary Visual Cortex." *Journal of Neuroscience* 17:7926–7940.

Varley, Thomas F., Olaf Sporns, Aina Puce, and John Beggs. 2020. "Differential Effects of Propofol and Ketamine on Critical Brain Dynamics." *bioRxiv.*

Vattay, Gábor, Dennis Salahub, István Csabai, Ali Nassimi, and Stuart A. Kaufmann. 2015. "Quantum Criticality at the Origin of Life." *Journal of Physics: Conference Series* 012023.

Veatch, Sarah L., Pietro Cicuta, Prabuddha Sengupta, Aurelia Honerkamp-Smith, David Holowka, and Barbara Baird. 2008. "Critical Fluctuations in Plasma Membrane Vesicles." *ACS Chemical Biology* 3:287–293.

Villegas, Pablo, José Ruiz-Franco, Jorge Hidalgo, and Miguel A. Muñoz. 2016. "Intrinsic Noise and Deviations from Criticality in Boolean Gene-Regulatory Networks." *Scientific Reports* 6:1–13.

Wagner, Hermann, Sandra Brill, Richard Kempter, and Catherine E. Carr. 2005. "Microsecond Precision of Phase Delay in the Auditory System of the Barn Owl." *Journal of Neurophysiology* 94:1655–1658.

Wang, Sheng-Jun, and Changsong Zhou. 2012. "Hierarchical Modular Structure Enhances the Robustness of Self-Organized Criticality in Neural Networks." *New Journal of Physics* 14:023005.

Weber, Ernst Heinrich. 1996. *E. H. Weber on the Tactile Senses.* New York: Psychology Press.

White, Edward L. 1989. *Cortical Circuits: Synaptic Organization of the Cerebral Cortex—Structure, Function, and Theory.* New York: Springer.

Williams-Garcia, R. V., M. Moore, J. M. Beggs, and G. Ortiz. 2014. "Quasicritical Brain Dynamics on a Nonequilibrium Widom Line." *Physical Review E—Statistical, Nonlinear, and Soft Matter Physics* 90:062714.

Wilson, Hugh R., and Jack D. Cowan. 1972. "Excitatory and Inhibitory Interactions in Localized Populations of Model Neurons." *Biophysical Journal* 12:1–24.

Wilson, Hugh R., and Jack D. Cowan. 1973. "A Mathematical Theory of the Functional Dynamics of Cortical and Thalamic Nervous Tissue." *Kybernetik* 13:55–80.

Wilson, Kenneth G. 1979. "Problems in Physics with Many Scales of Length." *Scientific American* 241:158–179.

Wilson, Matthew A., and Bruce L. McNaughton. 1994. "Reactivation of Hippocampal Ensemble Memories during Sleep." *Science* 265:676–679.

Wilting, Jens, and Viola Priesemann. 2018. "Inferring Collective Dynamical States from Widely Unobserved Systems." *Nature Communications* 9:1–7.

Wilting, Jens, and Viola Priesemann. 2019a. "25 Years of Criticality in Neuroscience—Established Results, Open Controversies, Novel Concepts." *Current Opinion in Neurobiology* 58:105–111.

Wilting, Jens, and Viola Priesemann. 2019b. "Between Perfectly Critical and Fully Irregular: A Reverberating Model Captures and Predicts Cortical Spike Propagation." *Cerebral Cortex* 29:2759–2770.

Wise, Steven P., and Elisabeth A. Murray. 2000. "Arbitrary Associations between Antecedents and Actions." *Trends in Neurosciences* 23:271–76.

Wong, Rachel O. L. 1999. "Retinal Waves and Visual System Development." *Annual Review of Neuroscience* 22:29–47.

Worrell, Gregory A., Stephen D. Cranstoun, Javier Echauz, and Brian Litt. 2002. "Evidence for Self-Organized Criticality in Human Epileptic Hippocampus." *Neuroreport* 13:2017–2021.

Wu, Jian-Young, Li Guan, and Yang Tsau. 1999. "Propagating Activation during Oscillations and Evoked Responses in Neocortical Slices." *Journal of Neuroscience* 19:5005–5015.

Wu, Jian-Young, Xiaoying Huang, and Chuan Zhang. 2008. "Propagating Waves of Activity in the Neocortex: What They Are, What They Do." *The Neuroscientist* 14:487–502.

Xu, Longzhou, Lianchun Yu, and Jianfeng Feng. 2021. "Avalanche Criticality in Individuals, Fluid Intelligence and Working Memory." *bioRxiv: 2020.08.24.260588.*

Yaghoubi, Mohammad, Ty de Graaf, Javier G. Orlandi, Fernando Girotto, Michael A. Colicos, and Jörn Davidsen. 2018. "Neuronal Avalanche Dynamics Indicates Different Universality Classes in Neuronal Cultures." *Scientific Reports* 8:1–11.

Yamins, Daniel L. K., and James J. DiCarlo. 2016. "Using Goal-Driven Deep Learning Models to Understand Sensory Cortex." *Nature Neuroscience* 19:356–365.

Yamins, Daniel L. K., Ha Hong, Charles F. Cadieu, Ethan A. Solomon, Darren Seibert, and James J. DiCarlo. 2014. "Performance-Optimized Hierarchical Models Predict Neural Responses in Higher Visual Cortex." *Proceedings of the National Academy of Sciences* 111:8619–8624.

Yeckel, Mark F., and Theodore W. Berger. 1990. "Feedforward Excitation of the Hippocampus by Afferents from the Entorhinal Cortex: Redefinition of the Role of the Trisynaptic Pathway." *Proceedings of the National Academy of Sciences* 87:5832–5836.

Young, George F., Luca Scardovi, Andrea Cavagna, Irene Giardina, and Naomi E. Leonard. 2013. "Starling Flock Networks Manage Uncertainty in Consensus at Low Cost." *PLoS Computational Biology* 9:e1002894.

Yu, Shan, Andreas Klaus, Hongdian Yang, and Dietmar Plenz. 2014. "Scale-Invariant Neuronal Avalanche Dynamics and the Cut-Off in Size Distributions." *PLoS One* 9:e99761.

Zack, M. M., and R. Kobau. 2017. "National and State Estimates of the Numbers of Adults and Children with Active Epilepsy—United States, 2015." *MMWR: Morbidity and Mortality Weekly Report* 66 (31): 821.

Zapperi, Stefano, Kent Bækgaard Lauritsen, and H. Eugene Stanley. 1995. "Self-Organized Branching Processes: Mean-Field Theory for Avalanches." *Physical Review Letters* 75:4071.

Zhang, Bintian, Weisi Song, Pei Pang, Yanan Zhao, Peiming Zhang, István Csabai, Gábor Vattay, and Stuart Lindsay. 2017. "Observation of Giant Conductance Fluctuations in a Protein." *Nano Futures* 1:035002.

Zhou, Z. Jimmy, and Dichen Zhao. 2000. "Coordinated Transitions in Neurotransmitter Systems for the Initiation and Propagation of Spontaneous Retinal Waves." *Journal of Neuroscience* 20:6570–6577.

Zimmern, Vincent. 2020. "Why Brain Criticality Is Clinically Relevant: A Scoping Review." *Frontiers in Neural Circuits* 14:54.

Index

Action potential (*also* Spike), 28, 32, 110, 119
Alpha band oscillations, 18
Arcangelis, Lucilla de, 167
Ashby, Ross, 109
Association cortex, 146, 149, 150–151, 161
Associations, 73, 149–150, 153–154, 156, 160
Autocorrelation, 18, 159, 161. *See also* Correlation
Avalanche, in iron, 96
Avalanche, in sandpiles, 94, 98, 134
Avalanche, neuronal, definitions of, 19, 54, 114
Avalanche, triggering, 136
Avalanche duration distribution, 19, 21
  when diverging, 169–170
  and exponent relation, 56–57, 66–68, 97, 99, 100, 174–175
  and homeostasis, 112, 116
  and human behavior, 128
  and quasicriticality, 133–134
  shifting of, 134
Avalanche examples, 15, 58, 104, 177, 178
Avalanche fragmenting, or splitting, 141
Avalanche merging, or concatenating, 135, 141
Avalanche patterns, 88–89, 177, 178. *See also* Repeating patterns
Avalanches, neuronal, 1, 5
  and branching ratio, 53
  and cortical layers, 156–158
  and epilepsy, 122
  experimentally discovered, 20–21
  explained, 14–15
  homeostasis of, 111, 115–118, 120
  and human behavior, 126–129
  and quasicriticality, 134–135
  and temporal correlations, 167
  universality of, 101–102, 105
Avalanches, runaway, 122, 140
Avalanches, scale-free property, 23, 56–57, 59, 66, 71, 175–177
Avalanche shape (or profile), 56–59, 63, 66, 68, 70–71, 100, 103–104, 126, 156
Avalanche shape collapse. *See also* Universal scaling function
  in cortical layers, 156–157
  in dissociated cultures, 63

examples of, 58–59, 68
as fractal copies, 66, 68–71
in memristor networks, 166
in non-critical dynamics, 143
how to perform, 176
software for, 180
in universality, 100–102
widely found, 163
Avalanches in nanowire networks, 166
Avalanches in vitro, 62–63
Avalanche size, temporal correlations, 167
Avalanche size distribution
  when diverging, 169–170
  experimentally first observed, 19–21
  and exponent relation, 56–57, 66–68, 103–104, 174–175
  and homeostasis, 112, 115–116, 118–119
  and human behavior, 128
  and Ising model, 96–99
  and kappa measure, 82–83
  and phase transition, 65–66
  and quasicriticality, 133–134
  and seizures, 122, 124–126
  as a signature of criticality, 25
  relation to spontaneous activity, 53
  and universality, 101–102
Avalanche size distribution, shifting of, 134, 141
Avalanche size distribution, subcritical, 65
Avalanche size distribution, supercritical, 65
Avalanche size vs duration, 56–58, 66–67, 97, 100, 112, 156–157, 174–175

Bak, Per, 3, 134, 166
Banach, Stefan, 145
Behavior, correlations with criticality, 126–129, 165, 173
Bertschinger, Nils, 89, 155
Bialek, Bill, 27, 166–167
Bonachela, Juan, 134, 139
Branching model
  as a branching process, 19–20, 124
  and branching ratio, 51–54
  and data, in agreement with, 66–67, 71, 83–87
  and dynamic range, 79